Geotours Workbook

A Guide for
Exploring Geology & Creating Projects
using Google Earth™

M. SCOTT WILKERSON
Department of Geosciences
DePauw University

M. BETH WILKERSON
Library & Information Services
DePauw University

STEPHEN MARSHAK
Department of Geology
University of Illinois

W. W. Norton & Company
New York • London

W. W. Norton & Company has been independent since its founding in 1923, when William Warder Norton and Mary D. Herter Norton first published lectures delivered at the People's Institute, the adult education division of New York City's Cooper Union. The Nortons soon expanded their program beyond the Institute, publishing books by celebrated academics from America and abroad. By mid-century, the two major pillars of Norton's publishing program—trade books and college texts—were firmly established. In the 1950s, the Norton family transferred control of the company to its employees, and today—with a staff of four hundred and a comparable number of trade, college, and professional titles published each year—W. W. Norton & Company stands as the largest and oldest publishing house owned wholly by its employees.

ISBN: 978-0-393-91891-5 (pbk.)

W. W. Norton & Company, Inc., 500 Fifth Avenue, New York, NY 10110
www.wwnorton.com

W. W. Norton & Company, Ltd., Castle House, 75/76 Wells Street, London W1T 3QT

4 5 6 7 8 9 0

Geotours Workbook

Section 2: Exploring Geology Using Google Earth

Section 3: Developing Your Own Interactive Google Earth Materials

Geotours Workbook

If you're reading this, you probably (a) already have spent numerous hours exploring the seductive virtual world of Google Earth and yearn to learn more about the software and the planet on which we live or (b) haven't yet used Google Earth (perhaps you're taking a course that requires you to use this workbook) and are on the brink of realizing how empty your life has been up to this point! In either case, sit back—your world is about to be "rocked."

Google Earth is a free "virtual globe" that simulates the planet on which we live by draping high-resolution global satellite imagery onto a sphere with 3-D terrain—over *the entire planet*. With a click of a mouse, users can take interactive virtual field trips anywhere in the world to explore the Earth's surface (even the ocean floor). Using the interactive 3-D interface, users literally can zoom and fly around to visualize any region on the planet from any direction, distance, and angle. As if that isn't enough, Google Earth also has the tools to be a "geographic web browser" to associate text, pictures, movies, etc. with specific spatial locations on the virtual globe. There's more—lots more, but you don't want to learn everything about Google Earth in the preface, do you?

This workbook started life back in 2008 (yes, books do take on a life of their own). What originated as a series of placemarks (aka Geotours) to automatically whisk users to various locations all over the planet to highlight cool geologic features, but it quickly morphed into a textbook supplement with worksheets to help students better understand what they were seeing. Predictably, with time (and new textbook editions), the Geotours and worksheets multiplied. Inevitably, users also wanted to know how to make their own Geotours and/or how to teach their students to create Geotour projects (hence, the workbook that you have before you).

You may be asking—*How should I use this book*? That depends on what you want to do...

All users
All users should read **Section 1**, which provides a basic introduction to (1) downloading and installing Google Earth and the files associated with this workbook and (2) using Google Earth.

Users who only want to learn about the Earth
Users who only want to learn about the Earth (e.g., you are taking an earth science or geology class) should focus on the Geotour worksheets in **Section 2** and the associated Geotour KMZ files. The worksheets are meant to supplement a geoscience textbook, so it is important to refer to such a textbook and/or similar geoscience reference materials (online or print) for background information in order to complete the worksheets.

For classroom use, worksheets typically are assigned as student homework or in-class assignments (see Table 1 for academic users who have adopted *Earth: Portrait of a Planet* or *Essentials of Geology* by Stephen Marshak). Alternatively, instructors may want to use materials in the Geotour KMZ files during lectures and/or to direct interested students to these materials for self-exploration.

Section 2 Geotour worksheet	Earth: Portrait of a Planet chapter	Essentials of Geology chapter
	Table 1. Using *Geotours Workbook* with *Earth: Portrait of a Planet* or *Essentials of Geology* by Stephen Marshak (W. W. Norton, Inc).	
A	1	1
B	2–4	2
C	5	3
D	6	4
E	9	5
F	7	6
G	8	7
H	10	8
I	11	9
J	12	10
K	13	11
L	14–15	12
M	16	13
N	17	14
O	18	15
P	19	16
Q	21	17
R	22	18
S	23	19

Users who only want to learn how to create Google Earth materials

Users who want to only learn how to create materials for Google Earth can go directly to **Section 3**. This section provides step-by-step instructions for creating Google Earth materials without droning on with extensive text explanations. As such, our approach is to divide each module into "projects" that generally contain less than 10 steps and have a single focused objective in mind. That is, in the Placemarks module, we create a placemark in one project, edit the placemark in a second project, format the text in a third project, add an image in a fourth project, etc...you get the idea.

We think this approach should work as well in the classroom with a wide range of student abilities as it does as a workbook for individuals who want to work through the material on their own. Examples are heavily geared toward geology, earth science, environmental science, etc., but the techniques are applicable to a wide array of disciplines.

<u>Users who want to learn about the Earth and then create their own geoscience-based Geotour projects or other Google Earth materials</u>
We like you! We would recommend that you at least work through some of the worksheets in **Section 2** before starting **Section 3** (the more of Section 2, the better). That way, you not only obtain some exposure to geoscience topics in the context of Google Earth, but you also gain an appreciation for the range of Google Earth's capabilities.

We've used this approach in the classroom, and it works quite well. Typically, we assign topics in **Section 2** throughout the semester, and then ask students to complete a Geotour project as a final semester project using **Section 3**. We typically work through **Section 3** as a group during a couple of lab periods (although many students are quite capable of working through the projects on their own).

Our goal is that this workbook helps you to explore important interdisciplinary geoscience topics in their proper spatial context, to better recognize geologic landforms and environmental issues affecting our planet, and to attain a basic knowledge of how these features and issues have formed and evolved. In addition, we hope that the workbook helps you develop the technical skills you need to develop your own Google Earth materials for your own projects. Enjoy!

Scott, Beth, & Steve

M. Scott Wilkerson
Professor of Geosciences
DePauw University, Department of Geosciences
Julian 217, 602 South College Avenue, Greencastle, IN 46135
mswilke@depauw.edu

M. Beth Wilkerson,
GIS Specialist
DePauw University, Library & Information Services
Julian 110, 602 South College Avenue, Greencastle, IN 46135
bwilkerson@depauw.edu

Stephen Marshak
Professor of Geology and Director of the School of Earth, Society, & Environment
University of Illinois, Department of Geology
208 Natural History Building, 1301 W. Green St., Urbana, IL 61801
smarshak@illinois.edu

We're very grateful for the assistance and support of numerous individuals and organizations during the development of this book. Carol Smith and Neal Abraham at DePauw University provided exceptional support and guidance for working on this project, especially during the early development phase of the Google Earth "how-to" materials and workshops. Jennifer Wasson of Precision Graphics helped during the earliest stages of this project by designing the layout of the original Geotour two-page spreads. Various individuals and organizations listed in the Credits and in the KMZ file were gracious to share photos, figures, diagrams, KML data, etc. with us. Jack Repcheck, Eric Svendsen, Rob Bellinger, Matthew Freeman, and many other W.W. Norton employees provided important logistical and editorial help, without which this book would never have been published. Lastly, Zach and Ben Wilkerson were excellent companions while Scott & Beth worked on this project, never complaining about the long hours at the office and always providing love and moral support.

Scott Wilkerson, Beth Wilkerson,
& Stephen Marshak

© 2011 Europa Technologies
US Dept of State Geographer
© 2011 Google
© 2011 MapLink/Tele Atlas

lat 37.529248° lon -94.621192° elev 257 m

Google earth

Eye alt 8391.75 km

Part 1

Downloading & Installing Files

In this part, you will learn how to download the Google Earth application and related Geotour Workbook materials.

Topic 1.1	**downloading the Google Earth application**

1. Using your web browser, go to the following website: http://earth.google.com/.

2. On this webpage, there is a button to download the free version of Google Earth. Click on this button to review the license agreement and to begin downloading (the site automatically recognizes your operating system and downloads the appropriate version for your computer). Alternatively, you can also choose to purchase Google Earth Pro (for a comparison of the two versions, go to http://www.google.com/enterprise/earthmaps/pro_features.html/).

3. Follow the on-screen instructions to install the Google Earth application on your particular system. The installer usually defaults to installing the application in the folder that contains all of your applications.

> *What version of Google Earth was used to create this workbook?*
>
> *This workbook was created using* **Google Earth 6.1***. Please note that the Google Earth application is constantly changing, and so interface elements, imagery quality, and layer content included with Google Earth may look somewhat different from what is shown in this workbook.*

Topic 1.2	**downloading Geotour Workbook files**

1. Using your web browser, download the following file to your desktop:
   http://media.wwnorton.com/college/geo/geotours/Geotours.kmz

2. Double-click the **Geotours.kmz** file on your desktop, and Google Earth will launch automatically (or select **File > Open > Geotours.kmz** from within Google Earth).

3. In the left-hand sidebar you will see a **Places** panel, and in the **Temporary Places** folder you will see a **Geotours.kmz** file. Drag the **Geotours.kmz** file into the **My Places** folder and then click on the triangle next to **Geotours.kmz** to reveal the **Geotours** subfolder.

4. Click on the triangle next to the **Geotours** folder to expand the folder and reveal its contents:
   Welcome,
   Section 1. *Getting Started*
   Section 2. *Geotours: Exploring Geology Using Google Earth*
   Section 3. *Developing Your Own Google Earth Geotours*

5. The **Welcome** folder contains several **Favorite Geotour Sites** that highlight Google Earth's capabilities as a means of better understanding the planet on which we live. This is the ideal place to begin your exploration of the Geotours.

6. The **Section 1** folder provides instructions on how to (1) download and install Google Earth and the files associated with this workbook, and (2) use Google Earth.

7. The **Section 2** folder contains two folders:

   a. The **Geotour Worksheets** folder contains files for each worksheet question in Section 2 of *Geotours Workbook*, arranged into folders by geoscience topics that correspond to chapters commonly used in most introductory geoscience textbooks.

   b. The **Geotour Site Library** provides a comprehensive library of all available Geotours associated with *Geotours Workbook* arranged into folders by geoscience topics that correspond to chapters commonly used in most introductory geoscience textbooks. Each of these folders contains

      i. a "Marshak-See For Yourself Sites" folder that contains placemarks for each of the locations shown in the Google Earth Appendix of *Earth: Portrait of a Planet* or *Essentials of Geology* by Stephen Marshak for academic users who have adopted those books (*please note that (1) these sites are useful for non-Marshak adopters as well, and (2) Marshak adopters can download a separate KMZ file with only the See For Yourself Sites from the W. W. Norton website*), and

      ii. Geotours or folders of grouped Geotours that pertain to the geoscience topic being studied. While totally independent of the Marshak-See For Yourself Sites and the Geotour Worksheets, these Geotours may duplicate and/or expand on sites visited in those two folders plus add many more sites related to the geoscience topic being studied.

8. The **Section 3** folder provides files illustrating how to develop Google Earth Geotour content.

9. Enjoy!

Part 2

Using Google Earth

In this part, you will learn how to use the Google Earth interface. In addition, you will learn about many of Google Earth's powerful features and the content that is provided free with the application. *Please note that SCROLL=mouse scroll wheel, LMB=left mouse button, and RMB=right mouse button. Unless specified, "click" refers to LMB.*

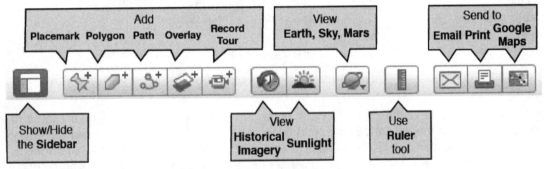

| Topic 2.1 | **the toolbar** |

When you start the Google Earth application, the program will initialize and an image of the Earth with a background of stars will appear on your screen. A **toolbar** across the top of the viewer window (Figure 2.1) provides **icons** that are quick links to performing various tasks.

Figure 2.1: Google Earth toolbar.

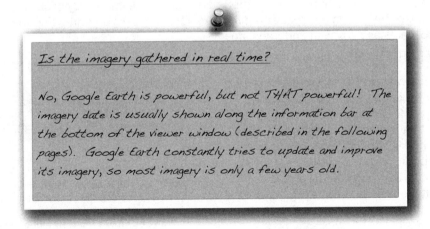

Is the imagery gathered in real time?

No, Google Earth is powerful, but not THAT powerful! The imagery date is usually shown along the information bar at the bottom of the viewer window (described in the following pages). Google Earth constantly tries to update and improve its imagery, so most imagery is only a few years old.

Topic 2.2   the navigation controls

In Google Earth, navigation tools appear in the upper right-hand corner of the screen (Figure 2.2).

The **Look tool** (circle with an eye in the center) allows you to rotate the view. You can do this either by clicking LMB and dragging the mouse on or within the outer ring, or by rotating the view clockwise or counterclockwise by clicking on the left or right arrows, respectively. Clicking (or dragging) the N

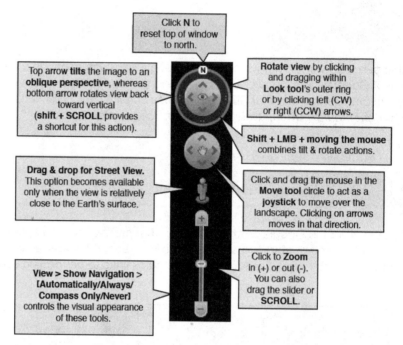

Figure 2.2: *Google Earth navigation controls.*
For additional information, see the Google Earth User Guide
(http://earth.google.com/support/bin/static.py?page=guide_toc.cs).

button with LMB reorients the view with north at the top of the screen. Clicking on the top arrow of the **Look tool** tilts the image to show an oblique view, whereas clicking on the bottom arrow rotates the perspective back toward vertical *(shift+SCROLL provides a shortcut for this action).* Clicking LMB and dragging the mouse in the inner circle of the **Look tool** performs both actions simultaneously *(shift+LMB+moving the mouse provides a shortcut for this action).*

The **Move tool** (circle with a hand in the center) pans across the image. By clicking the arrows, you move a finite distance in the specified direction. Clicking LMB and dragging the mouse in the inner circle acts as a joystick and translates the view in any direction.

As you zoom toward the ground surface, the **Street View tool** appears in the navigation controls (Figure 2.2, person icon). Dragging the person icon to a location in the Google Earth window will highlight nearby roads in blue if Street View photography is available. Placing the person icon along one of these roads will zoom and tilt the view to provide a ground-level perspective. The screen will enter Street View mode with Street View photography on the screen and information/controls in the upper righthand corner (e.g., the street address, a

Figure 2.3: *Google Earth Street View/ground-level view controls. Click on an object in Street View to advance to that object. RMB or Control-LMB followed by moving the cursor down-up will zoom in and out of the Street View imagery, respectively. SCROLL moves the view along the street.*

person button highlighted in blue, and the **Exit Street View button**; Figure 2.3). To toggle from Street View to ground-level view, click the building button next to the person button. The building button will highlight in blue, the address will disappear, and the Exit Street View button changes to **Exit ground-level view**. Note that the ground-level mode will appear if the person icon is placed at a location not associated with a road (or if the road has no Street View photography).

The **vertical Zoom slider** zooms the view up or down. You can either move the slider, use SCROLL, or click on the end of the bar. Clicking on the + end takes you to a lower elevation, whereas clicking on the – end takes you to a higher elevation. The slider offers better control. Select **View > Scale Legend** from the menu to display a bar scale in the lower left portion of the screen. The default setting in Google Earth is to automatically swoop (tilt) into a perspective view as you zoom closer. This auto-tilt behavior may be turned off
(*On a Mac*: *Google Earth > Preferences > Navigation > Navigation controls or*
On a PC: *Tools > Options > Navigation > Navigation controls*).

On the Earth image, you will see a hand-shaped cursor. By dragging the cursor across the screen, you can move the image. If you quickly drag the hand cursor while holding down the mouse and then let go, the movement will continue. Clicking anywhere in the viewer will stop this continuous motion of the Earth.

Topic 2.3   the information bar

On the bottom rule of a Google Earth window, you will see several information items (Figure 2.4).

On the left side of the screen, you will find a **scale bar** for the view (toggle on/off using **View > Scale Legend**), the **date of the imagery** in the window, and if zoomed in enough, a **Historical Imagery link** with the date of the oldest imagery available in Google Earth.

In the center of the screen, the location coordinates and ground surface elevation of the point just beneath the hand-shaped cursor on the screen are specified. The location coordinate units default to **latitude** (north [+] or south [-] of the equator) and **longitude** (east [+] or west [-] of the of the prime meridian). The coordinate system and format can be changed in the **Preferences (Mac)/Options (PC)** dialogs. Two versatile choices are decimal degrees (latitude/longitude) or Universal Transverse Mercator (a metric-based coordinate system referenced to latitude/longitude).

Finally, on the right side of the screen, the height of the **viewing elevation** ("Eye alt") is provided along with a **blue circle** that graphically shows if your computer is busy processing (e.g., streaming image data as you zoom in—at first the image is blurry, and then as streaming approaches 100%, it becomes clearer).

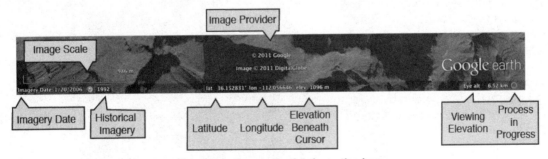

Figure 2.4: Google Earth information bar.

Topic 2.4   the panels sidebar

The sidebar contains information about locations and provides options for adding information to the screen image (Figure 2.5).

The topmost panel in the sidebar is the **Search panel**. There you can type a location (e.g., general name, specific address, latitude/longitude, etc.) in the text field and click the **Magnifying Glass** icon to execute the search (and then fly directly to the location). To remove the list of search items, click the blue "X" that appears in the lower right corner of the **Search panel**.

The middle area is the **Places panel** (Figure 2.5). It stores locations that you create yourself or that you open using **File > Open** (e.g., *Geotours Workbook* files). Clicking the box next to a name will toggle between showing/hiding the feature. Double-clicking the icon (recommended) or name of an item within a folder will take you to that location. A toolbar at the bottom of the **Places panel** allows you to search for objects, adjust an object's opacity, and take a tour of the object(s). Search by clicking the **Magnifying Glass** icon (Figure 2.5) and then entering text into the adjacent text field. Folders will automatically expand to show you items that meet your search criteria. The arrows next to the text field at the bottom of the **Places panel** will take you to the previous (↑) or next (↓) instance that matches your search. Unfortunately, folders remain expanded and must be closed manually.

The **Places panel** has two main folders: **My Places** and **Temporary Places**. Items in the **My Places** folder are automatically saved each time you exit Google Earth, and the previous **My Places** folder is saved as a backup (or *File > Save > Save My Places*). Look for myplaces.kml and myplaces.backup.kml on your system to locate these files (*On a Mac: Macintosh HD\ {username}\Library\Application Support\Google Earth\ or On a PC (XP): at C:\Documents and Settings\{username}\Application Data\Google\GoogleEarth\ or On a PC (Win 7): at C:\Users\ {username}\AppData\LocalLow\Google\GoogleEarth\ ).* If there is a corrupted **My Places** folder, you can delete the myplaces.kml folder, rename myplaces.backup.kml to myplaces.kml, and restart Google Earth. Items in the **Temporary Places** folder will not be automatically saved when you exit Google Earth. Upon exiting Google Earth, a dialog will ask you if you want to move the unsaved items to your **My Places** folder. If you click **Save**, those files will be available in your **My Places** folder in the **Places** panel the next time you open Google Earth. Alternatively, you can drag items from the **Temporary Places** folder to the **My Places** folder (and vice versa) at any time.

The lowermost panel in the sidebar is the **Layers panel**, which shows content provided by Google Earth and associated collaborators (Figure 2.5). For example, when you click on "Borders and Labels" in the **Layers** panel, country and state boundaries as well as city names will appear to provide a visual reference frame.

Any of these three panels (**Search**, **Places**, and **Layers**) may be hidden by toggling the triangle next to their respective name (Figure 2.5).

Why is it taking a long time to load my selected locations?

Checking an entire folder will cause all of its contents to load into memory at once, slowing down the system. It is better to be selective about which features you toggle on at the same time. Also, having a large My Places folder will slow down opening Google Earth.

6

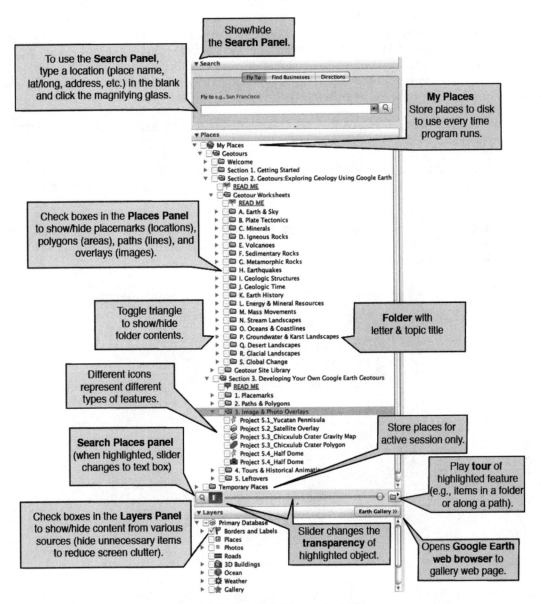

Show/hide the **Search Panel**.

To use the **Search Panel**, type a location (place name, lat/long, address, etc.) in the blank and click the magnifying glass.

My Places
Store places to disk to use every time program runs.

Check boxes in the **Places Panel** to show/hide placemarks (locations), polygons (areas), paths (lines), and overlays (images).

Toggle triangle to show/hide folder contents.

Folder with letter & topic title

Different icons represent different types of features.

Store places for active session only.

Search Places panel (when highlighted, slider changes to text box)

Play **tour** of highlighted feature (e.g., items in a folder or along a path).

Check boxes in the **Layers Panel** to show/hide content from various sources (hide unnecessary items to reduce screen clutter).

Slider changes the **transparency** of highlighted object.

Opens **Google Earth web browser** to gallery web page.

Figure 2.5: Google Earth panels sidebar.

7

Topic 2.5 — using the Search & Layers panels

The **Search** panel is your guide to traveling to all sorts of wonderful places: famous place names, specific addresses, even latitude and longitude coordinates. Once you click on the **Magnifying Glass** icon to the left of the search text field, you will be whisked away to your specified locale (or offered a series of links from which to choose). Once you arrive at your location, you can explore the surrounding area by toggling various layers on and off in the **Layers** panel. The **Layers** panel is loaded with all kinds of neat content that we encourage you to explore (be forewarned—you're probably going to learn something).

In this project, we're going to search for sites using different techniques and then examine these sites using layers.

1. Type khufu pyramid, giza in the **Fly to** blank of the **Search** panel. Press **Enter/Return** or click the **Magnifying Glass** icon to fly to this area. Usually one of the first search links will fly you directly there.

2. In the **Layers** panel, turn on **3-D Buildings > Photorealistic**. Use the Navigation controls to change your perspective to zoom into the area and to look west at the Sphinx and the three pyramids of Giza, Egypt from an oblique angle (Figure 2.6).

Figure 2.6: The Great Pyramids of Giza, Egypt.

3. As you can see, the Google database has general place names for famous locations and features. It also can search for specific addresses. Why don't you type in *your address* and see if you can locate your home? For example, type your address in the following format: 1600 Pennsylvania Ave NW, Washington, DC 20500.

4. After typing in and flying to your address, go ahead and type in the address in Step 3 and turn on **3-D Buildings > Photorealistic** in the **Layers** panel. You should be treated to a detailed version of the White House, complete with trees and a flag (Figure 2.7). It is well worth your time to explore the National Mall (e.g., the Capitol Building, the Washington Monument, the Lincoln Memorial, etc.) with the 3D Buildings turned on (and it will give you some good practice navigating within the Google Earth viewer).

Figure 2.7: A view from the South Lawn of the White House.

5. Turn on **Borders and Labels** in the **Layers** panel and type 36.092194, -112.118035 into the **Search** field (these numbers are "latitude, longitude"). You should fly west over the United States to Plateau Point in the Grand Canyon. If you didn't notice the various boundaries and place names, take a minute and zoom out to see them. Notice that the closer you are to the ground, the more detailed the information becomes.

6. Now turn on **More > Park and Recreation Areas**. As you zoom back to Plateau Point, notice the wealth of information that is now available to you, including feature names, park boundaries, trails, etc. Rotate your perspective so that you can see the topography of the Grand Canyon expressed in Google Earth with the layer information draped over the 3D surface (Figure 2.8). This 3D topography can be turned on and off by toggling the **Show Terrain** layer in the **Preferences (Mac)/Options (PC)** dialogs.

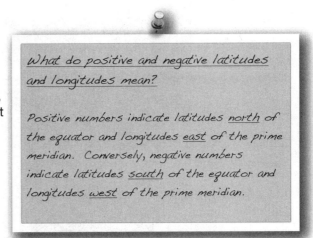

What do positive and negative latitudes and longitudes mean?

Positive numbers indicate latitudes north of the equator and longitudes east of the prime meridian. Conversely, negative numbers indicate latitudes south of the equator and longitudes west of the prime meridian.

Note: Google Earth in March 2010 removed the option to see more detailed labels (More > Geographic Features). Hopefully, these labels will be restored in response to user recommendations.

Figure 2.8: 3D view of the South Rim of the Grand Canyon.

7. Lastly, turn everything off in the **Layers** panel except **Borders and Labels** (*note: make sure Show Terrain is on in the **Preferences (Mac)/Options (PC)** dialogs*). Turn on **Gallery > Earthquakes** and **Gallery > Volcanoes**. Now zoom into the Aleutian Islands in Alaska. With these data turned on, you can readily see that this is a plate tectonic boundary that has abundant recent seismicity and active volcanism associated with it (Figure 2.9).

Figure 2.9: Earthquake and volcanic data near the Aleutian trench.

Topic 2.6   miscellaneous hints & tips

General
- If things look flat and lack a third dimension, check that **Show Terrain** is turned on in the **Preferences (Mac)/Options (PC)** dialogs.
- Some subtle features are more easily seen by temporarily increasing the vertical exaggeration to 3 *(Google Earth > Preferences (Mac) or Tools > Options (PC))*. A good general rule is to keep the vertical exaggeration at 1 (true elevation) or 1.5. A value of less than 1 flattens the terrain.
- Please be patient as images are loading into memory as some placemarks will look incomplete until the images are completely loaded.

Navigation
- You can turn off the auto-tilt feature:
  On a Mac: Google Earth > Preferences > Navigation > Navigation controls
  On a PC: Tools > Options > Navigation > Navigation controls
- It is easier to interact with Google Earth using a mouse (preferably with a scroll wheel) than with a trackpad.

Panels
- Do not turn on all the folders at once because the program will load everything at the same time. This is true not only for the **Places** panel, but also for the **Layers** panel (especially 3D buildings).
- You can make the Panels sidebar wider by clicking and dragging on the right side of the Panels sidebar (between it and the main Google Earth window).
- The **Search** and **Layers** panels can be hidden by clicking the triangle next to their name in order to see more of the **Places** panel.

Working with folders & files in the Places panel
- Folders and placemarks can only be moved one at a time (i.e., multiple selections are not allowed in Google Earth. It may be useful to create a new folder first *(Add > New Folder)*, highlight it, and then create new placemarks within it.
- Any folder or feature (e.g., placemark, polygon, overlay, etc.) can be saved as either a KML or KMZ file. To do so, right-click on the folder or feature in the **Places** panel and choose **Save Place As**. In the Save dialog box, choose either the KML or KMZ format and indicate where you want to save the file.

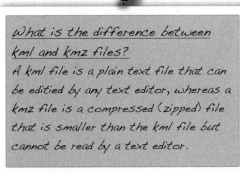

What is the difference between kml and kmz files?
A kml file is a plain text file that can be editied by any text editor, whereas a kmz file is a compressed (zipped) file that is smaller than the kml file but cannot be read by a text editor.

- Recall from Topic 2.4 that items in the **My Places** folder are automatically saved when you exit Google Earth, whereas items in the **Temporary Places** folder are not. Fortunately, when you exit Google Earth, a dialog will ask you if you want to move the unsaved items from your **Temporary Places** folder to your **My Places** folder before closing. Alternatively, you can drag items from the **Temporary Places** folder to the **My Places** folder (and vice versa) at any time.

Exploring Geology Using Google Earth

M. Scott Wilkerson, M. Beth Wilkerson, & Stephen Marshak

© 2011 Europa Technologies
US Dept of State Geographer
© 2011 Google
© 2011 MapLink/Tele Atlas

Google earth

lat 37.529248° lon -94.621192° elev 257 m

Eye alt 8391.75 km

Geotour Worksheet **A**

Earth & Sky

Image © 2007 DSS Consortium
Image NASA/STScI

Google earth

RA 5h34m31.97s Dec 22°00'52.08"   0°09'02.42" arcdegrees

To answer questions for this worksheet, go to the following Geotours folder in Google Earth:
2. Exploring Geology Using Google Earth > Geotour Worksheets > A. Earth & Sky

Supplemental background material also may be available in your textbook, through various internet resources, and within files in the 2. Exploring Geology Using Google Earth > Geotour Site Library.

A. Earth & Sky Worksheet

Nebular Supernova
- Crab Nebula

Spiral Galaxy
- Milky Way analog galaxy

Meteorite Impacts
- Manicouagan Crater, Canada
- Upheaval Dome, UT
- Meteor Crater, AZ
- Chesapeake Bay Crater, MD

Transition from Continental Crust to Oceanic Crust
- East coast, USA
- Eastern Greenland

Topography of the Ocean Floor
- Pacific Ocean

Nebular Supernova

1. Select *Google Sky* by clicking ![icon] and then selecting *Sky*. Turn on *Sky Database >
Imagery* in the Layers panel and double-click placemark Problem 1 in the **A. Earth & Sky**
worksheet folder. This placemark will take you to the spectacular Crab Nebula in *Google Sky*.
(a) The Crab Nebula is located about 6000 light years from Earth, and the present length of its
long axis is approximately 11 light years (a **light year** is the distance light travels in one
year...about 9.5 trillion km). What is the length of the long axis of the Crab Nebula (in km)?
- [ ] ~57,000 trillion km
- [ ] ~104.5 trillion km
- [ ] ~9.5 trillion km
- [ ] ~66,000 trillion km
(b) The Crab Nebula represents a violent **supernova explosion** of a once massive star, which
was likely a third, fourth, or even later generation star. Heavy elements with atomic weights
larger than 26 were likely generated during this explosion. Which element probably formed in
a supernova? *You may want to refer to your textbook or to a periodic table.*
- [ ] Gold (Au)
- [ ] Carbon (C)
- [ ] Hydrogen (H)
- [ ] Helium (He)

Spiral Galaxy

2. Double-click placemark Problem 2 while still in *Google Sky*. This image is of a spiral galaxy
that resembles what our Milky Way might look like if viewed from outside the galaxy. Note how
the curved spiral arms develop around the more quickly rotating central cluster of stars.
(a) Looking at the spiral arms <u>from this viewpoint</u>, in what direction is this galaxy rotating?
- [ ] counter-clockwise
- [ ] clockwise

Meteorite Impacts

3. Switch back to *Google Earth* by clicking [icon] and then selecting Earth. Now we'll visit four impact sites on Earth.

(a) Check and double-click the placemarks for Problem 3a. Use the Ruler Tool [icon] to determine the present-day diameter of Manicouagan Crater between the placemarks.
☐ ~5 km
☐ ~1.2 km
☐ ~96 km
☐ ~75 km

(b) Check and double-click the placemarks for Problem 3b. Use the Ruler Tool to determine the present-day diameter of Upheaval Dome between the placemarks.
☐ ~5 km
☐ ~1.2 km
☐ ~96 km
☐ ~75 km

(c) Check and double-click the placemarks for Problem 3c. Use the Ruler Tool to determine the present-day diameter of Meteor Crater between the placemarks.
☐ ~5 km
☐ ~1.2 km
☐ ~96 km
☐ ~75 km

(d) Check and double-click the placemarks for Problem 3d. Use the Ruler Tool to determine the present-day diameter of Chesapeake Bay Crater between the placemarks.
☐ ~5 km
☐ ~1.2 km
☐ ~96 km
☐ ~75 km

(e) Assume that a 40-m diameter meteorite created Meteor Crater. Although clearly an over-simplification, use a simple ratio between meteorite size and crater diameter to estimate the size of meteorite that might have created Manicouagan Crater (*use the crater diameters measured for Problems 3a & 3c*).
☐ ~40 m
☐ ~170 m
☐ ~2500 m
☐ ~3500 m

(f) On your computer, go to the Earth Impact Effects Program website at: http://www.lpl.arizona.edu/impacteffects/ This site estimates the consequences of an impact as a function of various parameters, including the size, velocity, and composition of the meteorite. Enter the following:

> *Distance from Impact: 1000 km*
> *Projectile Diameter: Manicouagan's estimated meteorite diameter from above (in m)*
> *Projectile Density: Trial 1-ice (comet) and Trial 2-iron (some asteroids)*
> *Impact Velocity: 20 km/s*
> *Impact Angle: 45 degrees*
> *Target Type: Crystalline Rock*

Perform Trials 1 & 2 to investigate the "impact" of changing projectile density. Which of the following statements is true?

- ☐ varying projectile density does not influence the final crater size
- ☐ icy comets would always melt before impacting the earth and creating a crater
- ☐ an icy comet produces a larger final crater than an iron asteroid
- ☐ an icy comet produces a smaller final crater than an iron asteroid

Just for Fun...*Use the Ruler Tool* *and the four crater diameters to see the consequence of what similar-sized impacts would be in the area in which you live!*

Transition from Continental Crust to Oceanic Crust

4. The transition from continental crust to oceanic crust produces distinctive features on the Earth's surface. Specifically, the **continental shelf** is the submerged edge of a continent, the **continental slope** is the relatively steep seaward edge of the shelf, and the **abyssal plain** is the deep, broad, low-relief area of the ocean floor. This transition is visible on *Google Earth* imagery.

(a) Check and double-click the placemarks labeled Problem 4a-i, -ii, and -iii off the east coast of the United States and identify which of the following is correct.
- ☐ 4a-i: shelf; 4a-ii: slope; and 4a-iii: abyssal
- ☐ 4a-i: shelf; 4a-ii: abyssal; and 4a-iii: slope
- ☐ 4a-i: slope; 4a-ii: abyssal; and 4a-iii: shelf
- ☐ 4a-i: abyssal; 4a-ii: shelf; and 4a-iii: slope

(b) Identify the area depicted by placemark Problem 4b near Eastern Greenland.
- ☐ continental shelf
- ☐ continental slope
- ☐ abyssal plain
- ☐ mid-ocean ridge

Topography of the Ocean Floor

5. The ocean floor is not all a featureless abyssal plain! In fact, it has features ranging from deep trenches to large mountain ranges.

(a) Check and double-click the placemarks labeled Problem 5a-i, -ii, -iii, and -iv and identify which of the following is <u>incorrect</u> (*you may need to refer to your textbook glossary*).
- ☐ 5a-i is a deep ocean trench
- ☐ 5a-ii is a seamount on the ocean floor
- ☐ 5a-iii is a mid-ocean ridge
- ☐ 5a-iv is a deep ocean trench

Plate Tectonics

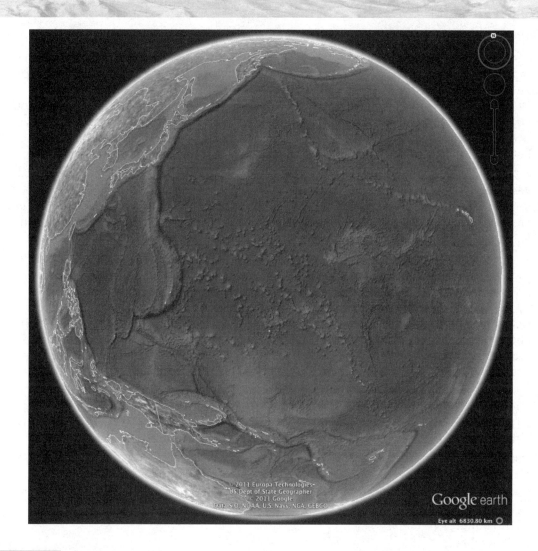

To answer questions for this worksheet, go to the following Geotours folder in Google Earth:
2. Exploring Geology Using Google Earth > Geotour Worksheets > B. Plate Tectonics

Supplemental background material also may be available in your textbook, through various internet resources, and within files in the 2. Exploring Geology Using Google Earth > Geotour Site Library.

B. Plate

Tectonics

Worksheet

Continental Drift
- Atlantic Ocean

Divergent Boundaries
- Mid-Atlantic Ridge
- East Pacific Rise
- East Africa triple junction

Convergent Boundaries
- Mariana Trench
- Tonga Trench
- Trench off west coast of South America

Transform Boundaries
- Mid-Atlantic Ridge
- San Andreas Fault, CA

Hot Spots
- Hawaiian-Emperor seamount hot-spot chain

Plate Tectonics

Continental Drift

1. Check the box next to "Seafloor Age Map" in the **B. Plate Tectonics** worksheet folder to view the ages of the oceanic crust around the world. *Note that you can use the transparency slider to make this overlay semi-transparent.*

(a) Check and double-click the placemarks for Problem 1 to fly to a location above the Atlantic Ocean. These placemarks represent **conjugate points** (locations on the opposite sides of an ocean that were once adjacent before **seafloor spreading** occurred). Although the fit between the African and South American coastlines had been recognized for some time,

Wegener carefully matched continental shelves to improve this fit. Use the Ruler Tool to determine how far these points have moved apart (in km). Zoom in and use the Path tab to create segments along the major fracture zone that offsets the colored ages of the ocean floor (round to the nearest 1000 km).

☐ 7000 km
☐ 2000 km
☐ 5000 km
☐ 3000 km

(b) Using the Seafloor Age Map, about how many millions of years ago (Ma) were these points once adjacent? *Note that part of the seafloor adjacent to the coast is not covered by the Seafloor Age Map. Toggle the map on/off to determine the colored age band nearest the coast.*

☐ 110-120 Ma
☐ 170-180 Ma
☐ 50-60 Ma
☐ 30-40 Ma

(c) Using the largest number of the range for your answer to 1b and using the distance for your answer to 1a, calculate the **average spreading rate** for these points in km/Ma.

☐ 58 km/Ma
☐ 42 km/Ma
☐ 83 km/Ma
☐ 75 km/Ma

(d) Express your average spreading rate answer for 1c in terms of cm/yr. *Note: If the plates are moving apart symmetrically at the same rate (i.e., the color band widths are approximately equal), then 1/2 of this answer is the average rate at which the South American plate is moving west and the African plate is moving east.*

☐ 7.5 cm/yr
☐ 5.8 cm/yr
☐ 42 cm/yr
☐ 4.2 cm/yr

Just for Fun...*Visit the region between Australia and Antarctica and see if you can determine conjugate points that were once joined. It is easier if the Seafloor Age Map is turned on!*

Divergent Boundaries

2. Check the box next to "Seafloor Age Map" to view the ages of the oceanic crust around the world. Make the map semi-transparent.

(a) The placemarks for Problem 2a lie on the crests of mid-ocean ridges (**divergent boundaries**) in the Atlantic and Pacific Oceans, respectively. Double-click each placemark and use the width of the color bands representing seafloor ages to determine which divergent boundary is spreading at a faster rate.
- ☐ Problem 2a-Atlantic
- ☐ Problem 2a-Pacific
- ☐ both are spreading at approximately the same rate

(b) Some divergent boundaries spread at different rates. Double-click on the Problem 2b placemark to fly above the East Pacific Rise. Use the widths of the color bands (dark red orange in particular) representing seafloor ages on either side of this mid-ocean ridge to determine which side is moving faster.
- ☐ east side
- ☐ west side
- ☐ both are spreading at approximately the same rate

(c) In the Layers panel, turn on *Borders and Labels* and also *Gallery > Volcanoes*, then double-click on placemark Problem 2c. Divergent boundaries often begin as **triple junctions** comprised of three "arms" at ~120° apart that may or may not evolve into divergent boundaries. Which of the statements below about this area is <u>incorrect</u>?
- ☐ the NW-trending Red Sea is a linear sea that is opening in a NE-SW direction
- ☐ the ENE-trending Gulf of Aden is a linear sea that is opening in a NNW-SSE direction
- ☐ the volcanoes in Africa define a narrow linear rift that is opening in a NW-SE direction
- ☐ the volcanoes in Africa define a volcanic arc related to subduction off Africa's east coast

Just for Fun...*Increase your vertical exaggeration to 3 (<u>On a Mac</u>: Google Earth > Preferences > 3D View > Elevation Exaggeration or <u>On a PC</u>: Tools > Options > 3D View > Elevation Exaggeration), zoom down to the line of volcanoes in Africa, and tilt your view. Fly over this line of volcanoes (known as the East African Rift) to view the landscape. Think about how this region may evolve similarly to Madagascar.*

Convergent Boundaries

3. Check the boxes next to "Seafloor Age Map" and "Earthquakes" folders to view the ages of the oceanic crust around the world and selected 1986-2005 earthquakes color-coded by depth, respectively. Make the map semi-transparent. Also, in the Layers panel, turn on *Borders and Labels* and *Gallery > Volcanoes*.

(a) Check and double-click placemark Problem 3a to fly to the Mariana Trench. Which of the following is <u>incorrect</u> (*think about where volcanoes form in association with a subduction zone*)?
- ☐ subducting plate is to the east and overriding plate is to the west
- ☐ subducting plate is to the west and overriding plate is to the east
- ☐ the line of volcanoes west of the Mariana Trench is the volcanic arc
- ☐ earthquakes become deeper to the west

(b) Subduction ultimately produces a volcanic arc on the overriding plate (placemark Problem 3b). What is the depth of the majority of earthquakes <u>directly</u> beneath the volcanic arc associated with the Mariana Trench?
- ☐ 0-50 km
- ☐ 51-100 km
- ☐ 101-200 km
- ☐ 201-400 km

(c) Use the Ruler Tool [icon] to determine the **arc-trench gap** (distance between the volcanic arc and the trench axis, in km) between the placemarks of Problem 3c.
- ☐ ~106 km
- ☐ ~137 km
- ☐ ~212 km
- ☐ ~433 km

(d) Check and double-click placemark Problem 3d to fly to the Tonga Trench. What is the depth of the majority of earthquakes <u>directly</u> beneath the volcanic arc associated with the Tonga Trench?
- ☐ 0-50 km
- ☐ 51-100 km
- ☐ 101-200 km
- ☐ 201-400 km

(e) Use the Ruler Tool [icon] to determine the **arc-trench gap** (in km) between the placemarks of Problem 3e.
- ☐ ~245 km
- ☐ ~401 km
- ☐ ~303 km
- ☐ ~190 km

(f) Check and double-click placemark Problem 3f to fly to South America. What is the depth of the majority of earthquakes <u>directly</u> beneath the volcanic arc associated with this subduction zone?
- ☐ 0-50 km
- ☐ 51-100 km
- ☐ 101-200 km
- ☐ 201-400 km

(g) Assume that the earthquake depths define the **Wadati-Benioff zone** for the subducting slab. Compare your answers to Problem 3b, 3d, and 3f and choose the statement that best describes your observations.
- ☐ subducting slabs must reach a depth of 51-100 km before they produce volcanic arcs
- ☐ subducting slabs must reach a depth of 101-200 km before they produce volcanic arcs
- ☐ subducting slabs must reach a depth of 201-400 km before they produce volcanic arcs
- ☐ volcanic arcs don't seem to have a systematic relationship with the subducting slab

(h) Given your answer to 3e for the Tonga Trench and your answer for 3g (use an average), calculate the angle of descent for the subducting slab using the following formula:
angle of slab descent = tan^{-1} (depth of slab beneath arc ÷ arc-trench gap).
- ☐ ~21°
- ☐ ~63°
- ☐ ~38°
- ☐ ~05°

Transform Boundaries

4. Check the boxes next to "Seafloor Age Map" to view the ages of the oceanic crust around the world. Make the map semi-transparent. Also, in the Layers panel, turn on *Borders and Labels* as well as *Gallery > Earthquakes*.

(a) Check and double-click the placemarks for Problems 4a-i, -ii, and -iii to fly to the Mid-Atlantic Ridge. These placemarks either point to a segment of an active **transform fault** or of an inactive **fracture zone** (*see the transform fault section of your textbook*). Which of the following statements is correct?

☐ 4a-i and 4a-iii are inactive fracture zones and 4a-ii is an active transform fault
☐ 4a-i and 4a-iii are active transform faults and 4a-ii is an inactive fracture zone
☐ 4a-i and 4a-ii are inactive fracture zones and 4a-iii is an active transform fault
☐ all three are along an active transform fault

(b) Which direction are the placemarks for Problem 4b-i and 4b-ii moving, respectively? *Think about the process of rifting at the mid-ocean ridges.*

☐ east, west
☐ west, east
☐ both east
☐ both west

(c) Which direction are the placemarks for Problem 4c-i and 4c-ii moving. respectively? *Think about the process of rifting at the mid-ocean ridges.*

☐ east, west
☐ west, east
☐ both east
☐ both west

(d) Check and double-click the placemark for Problem 4d. Given the offset stream shown here, what type of transform fault is the San Andreas Fault?

☐ right-lateral *(at the fault, you turn right to find the offset stream segment)*
☐ left-lateral *(at the fault, you turn left to find the offset stream segment)*

Hot Spots

5. Check the boxes next to "Seafloor Age Map" to view the ages of the oceanic crust around the world. Make the map semi-transparent. Also, in the Layers panel, turn on *Borders and Labels* as well as *Gallery > Volcanoes*.

(a) Check and double-click the folder Problem 5 to fly to the Pacific Ocean where you will have a view of the Hawaiian-Emperor seamount hot-spot chain with their age of formation in millions of years (Ma). Use the Ruler Tool to measure the distance between Midway Atoll and Kilauea (cm) and calculate the average velocity of the Pacific Plate in cm/yr.

☐ ~1.9 cm/yr
☐ ~4.1 cm/yr
☐ ~8.8 cm/yr
☐ ~15 cm/yr

(b) What is the approximate age of the bend in the Hawaiian-Emperor seamount hot-spot chain (*should you use the age of the volcanic features or the age of the seafloor*)?

☐ 120 Ma
☐ 150 Ma
☐ 32 Ma
☐ 43 Ma

Geotour Worksheet C

Minerals

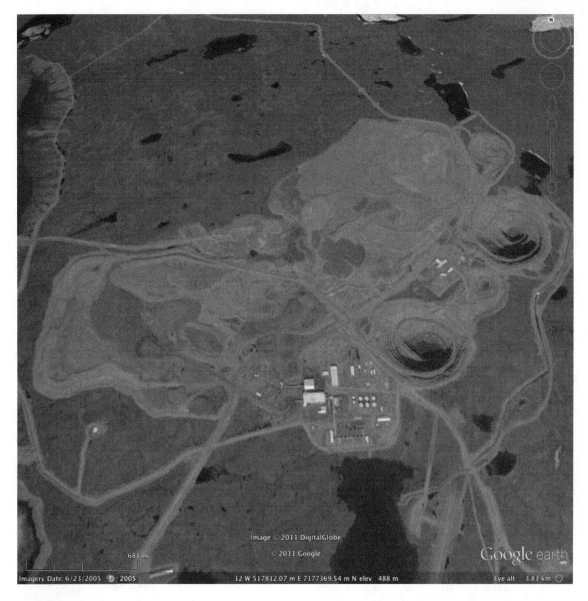

C. Minerals

Worksheet

To answer questions for this worksheet, go to the following Geotours folder in Google Earth:
2. Exploring Geology Using Google Earth > Geotour Worksheets > C. Minerals

Supplemental background material also may be available in your textbook, through various internet resources, and within files in the 2. Exploring Geology Using Google Earth > Geotour Site Library.

Diamond Mines
- Kimberley, South Africa

Mineral Reactions after Coal Mining
- Reclaimed coal mine, IN

Diamond Mining

1. We are all very familiar with the beautiful diamonds that adorn jewelry. However, the process by which some of the diamonds are mined is not quite so "beautiful". Type Kimberley, South Africa in the Search panel to fly to a location above Kimberley, South Africa (a region noted for its diamond mines). Zoom closer and explore this area a bit.

(a) Check and double-click the placemarks for Problems 1a-i, -ii, and -iii in the **C. Minerals** worksheet folder. Here, we see some of the nastier parts of the diamond mining process. Which feature is misidentified?

☐ 1a-i is a depression created by removing material during the mining process

☐ 1a-ii is excavated waste material (tailings) from mining operations piled on the land surface

☐ 1a-iii is an oddly colored lake likely contaminated by runoff from nearby tailings piles

☐ all features are correctly identified

Mineral Reactions after Coal Mining

2. The Kimberley example in Problem 1a may have made you question why the area has not been reclaimed or remediated. We agree wholeheartedly. Sometimes, however, even the best attempts at remediating former mining sites can face challenges.

(a) Check and double-click the placemarks for Problems 2a-i, -ii, and -iii. The area around placemark Problem 2a-i was a former coal strip mine that has been reclaimed. However, minerals like pyrite (iron sulfide-FeS_2) are commonly associated with coal and have been brought to the surface during the mining process where they can react and chemically weather more readily. Specifically, sulfuric acid and iron oxide (orange staining) are produced. Zoom in on areas 2a-ii and 2a-iii, then watch the Problem 2a-iii flyover and choose the incorrect statement.

☐ the remediation process has been successful in containing waste runoff at the mine site

☐ acidic, iron-rich water has traveled from the reclaimed mine into a nearby pond

☐ acidic, iron-rich water has flowed downstream in the creek, contaminating a larger area

☐ additional remediation must be done in order to contain the acid-mine drainage

Geotour Worksheet D
Igneous Rocks

D. Igneous Rocks Worksheet

To answer questions for this worksheet, go to the following Geotours folder in Google Earth:
2. Exploring Geology Using Google Earth > Geotour Worksheets > D. Igneous Rocks

Supplemental background material also may be available in your textbook, through various internet resources, and within files in the 2. Exploring Geology Using Google Earth > Geotour Site Library.

Batholiths and Laccoliths
- Yosemite National Park, CA
- Sierra Nevada Mountains, CA
- Henry Mountains, UT
- Mt. Hillers, UT

Dikes
- Spanish Peaks, CO
- Shiprock, NM

Igneous Rocks

Batholiths and Laccoliths

1. Yosemite National Park provides spectacular glaciated exposures of the Sierra Nevada **batholith**. The batholith is comprised of many separate plutons, which intruded into the area as part of a Mesozoic-age subduction system.

(a) Check and double-click the placemark for Problem 1a in the **D. Igneous Rocks** worksheet folder to fly to the spectacular Half Dome in Yosemite National Park. Given that it is part of an <u>intrusive</u> batholith, what type of rock texture would you expect to see at Half Dome?
- ☐ fine-grained
- ☐ coarse-grained
- ☐ glassy
- ☐ fragmental

(b) Visitors commonly note the rounded appearance of the back side of Half Dome. It was not polished by glaciers, but rather fractured off in sheets much like how one might peel layers from an onion. This process occurs because these rocks formed at ~20-25 km depth, but have experienced a removal of confining pressure since they are now exposed at the surface (think of how a sponge expands once you remove your hands pressing down on it). What is this process called? (*Hint: Look these words up in your glossary*).
- ☐ stoping
- ☐ assimilation
- ☐ fractional crystallization
- ☐ exfoliation

(c) Check and double-click on the "North America Batholiths map" in the folder labeled Problem 1. You will see the exposed batholiths of the western United States. Using the transparency slider at the bottom of the Places panel, make the overlay semi-transparent. To get a sense of the size of the Sierra Nevada batholith, determine its length between placemarks for Problem 1c using the Ruler Tool [image] (in km).
- ☐ 100-130 km
- ☐ 300-340 km
- ☐ 700-720 km
- ☐ 1000-1030 km

(d) Check and double-click on the red polygon for the Henry Mountains **laccolith** complex in the folder labeled Problem 1. This complex consists of many blister-like intrusions. Note the size difference between this intrusion and the Sierra Nevada batholith. Estimate how much smaller in length the Henry Mountains laccolith complex is relative to the batholith.
- ☐ ~1/2
- ☐ ~1/4
- ☐ ~1/6
- ☐ ~1/17

(e) Turn off the North America Batholiths map and the Henry Mts Laccolith Complex polygon and then double-click on the placemark for Problem 1e to zoom down to the south flank of the Mt. Hillers laccolith. The laccolith is comprised of gray andesite that has intruded into Mesozoic-age sedimentary rocks and upturned them on the flanks of the intrusion (recall that **laccoliths** are concordant intrusions that cause the overlying sedimentary layers to fold up in a blister or mushroom-shaped pattern). The sedimentary rocks over the intrusion have been eroded. Placemark Problem 1e highlights a tabular intrusion from the main laccolith that is parallel to the upturned sedimentary layers. What kind of intrusion is this?

☐ sill *(a tabular intrusion that inserts between sedimentary layers)*
☐ dike *(a tabular intrusion that cuts across sedimentary layers)*

Dikes

2. A **dike** is a tabular igneous intrusion that is discordant and cuts across pre-existing layering.

(a) Check and double-click placemark Problem 2a to fly to the Spanish Peaks, CO area where some prominent volcanic centers are visible. Which of the following statements is <u>incorrect</u>?

☐ igneous rock comprising the dike is more resistant to erosion than the surrounding rock
☐ there are no similar dikes in the surrounding area
☐ the dike is very linear, implying that magma likely filled a crack
☐ the dike's magma source is likely related to the Spanish Peaks volcanic center

(b) Check and double-click the placemarks for Problems 2b-i, -ii, and -iii. These placemarks highlight dikes filling radial fractures from the volcanic neck at Shiprock. Which of the following is the most reasonable explanation for this pattern?

☐ they filled fractures that existed in the area before the volcano developed
☐ the volcanic magma chamber created pressures that fractured the rocks in a radial pattern
☐ this fracture pattern is accidental and random
☐ the fracture pattern is controlled by the nearly horizontal sedimentary rock layers

Just for Fun...*Check and double-click on the polygon for Stone Mt., GA to fly to that location.*

Use the Ruler Tool [▮] *to measure the long axis of the* **stock** *that comprises Stone Mt. Clearly, this stock is much smaller than the Sierra Nevada batholith represented on the North America Batholiths map (recall that stocks often are the exposed tips of larger subsurface plutons like batholiths). In the future, given enough additional time and erosion, this area may be considered a batholith as well!*

To answer questions for this worksheet, go to the following Geotours folder in Google Earth:
2. Exploring Geology Using Google Earth > Geotour Worksheets > E. Volcanoes

Supplemental background material also may be available in your textbook, through various internet resources, and within files in the 2. Exploring Geology Using Google Earth > Geotour Site Library.

E. Volcanoes

Worksheet

Shield Volcanoes
- Mauna Loa, HI
- Kilauea, HI
- Mauna Kea, HI

Composite Cone Volcanoes
- Mt. St. Helens, WA
- Mt. Etna, Italy
- Mt. Vesuvius, Italy
- Crater Lake, OR

Cinder Cone Volcanoes
- SP Mountain, AZ
- Menan Buttes, ID
- Sunset Crater, AZ

Name:_____

Volcanoes

Shield Volcanoes

1. **Shield volcanoes** have a characteristic shape like a concave-down warrior's shield, in part because they are predominantly composed of low viscosity basaltic lava.

(a) Check and double-click the placemarks for Problem 1a in the Problem 1 folder within the **E. Volcanoes** worksheet folder to fly to Mauna Loa on the Big Island of Hawaii in the Hawaiian Islands. Turn on the "Mauna Loa contour map" in the Problem 1 folder. The brown lines are

contours (lines of equal elevations). Use the Ruler Tool [ ] to determine the **horizontal distance** between the placemarks for Problem 1a (*use feet as your units*) and then subtract the elevations (ft) for each placemark from the contour map to determine the **relief**. Find the slope of Mauna Loa using the following formula (this value is representative for many shield volcanoes): ***slope angle = tan⁻¹ (relief ÷ horizontal distance).***

$\square$ 5-10°
$\square$ 15-20°
$\square$ 25-35°
$\square$ 35-45°

(b) Check and double-click the placemarks for Problems 1b-i, -ii, and -iii. The basaltic lava flows weather according to their age (darker is younger). Place the three lava flows in chronologic order from youngest to oldest.

$\square$ Problem 1b-i, 1b-ii, and 1b-iii
$\square$ Problem 1b-ii, 1b-iii, and 1b-i
$\square$ Problem 1b-iii, 1b-ii, and 1b-i
$\square$ Problem 1b-ii, 1b-i, and 1b-iii

(c) Check and double-click placemark Problem 1c. Here, the fluid basaltic lava flows have flowed from Kilauea's eruptive vents over a steep cliff on their way to the ocean. Such a cliff is called a **pali**. Check and double-click on the "Hawaiian Island overlay" for a hint as to what the pali represents.

$\square$ the uppermost part of a landslide where the land has detached and moved to the SE
$\square$ an edge of a previous lava flow
$\square$ the marginal flank of an offshore caldera
$\square$ the erosional remnant of a former river bank

(d) Check and double-click placemark Problem 1d. Even from this distance, you can see the fluid, ropey flow patterns of this distinctive type of basaltic lava flow. What is it called?

$\square$ pahoehoe
$\square$ a'a'

(e) Check and double-click placemark Problem 1e to fly to the summit of Mauna Kea. A series of small parasitic volcanoes dot the landscape. These types of volcanoes often erupt during the latter stages of activity for the larger volcano and are comprised of mostly pyroclastic material. What type of volcanoes are they?

$\square$ shield
$\square$ composite cone
$\square$ stratovolcano
$\square$ cinder cone

Composite Cone Volcanoes

2. **Composite cones** (or **stratovolcanoes**) have the characteristic symmetrical shape most commonly associated with volcanoes. They are typically comprised of layers of andesitic lava and pyroclastic materials. Because of their composition, these volcanoes tend to be very explosive.

(a) Check and double-click the placemarks for Problem 2a to fly to Mt. Saint Helens in the Cascade Range of western Washington. Turn on the "Mt. Saint Helens contour map" in the Problem 2 folder. The brown lines are **contours** (lines of equal elevations). Use the Ruler Tool [tool] to determine the **horizontal distance** between placemarks for Problem 2a (*use feet as your units*) and then subtract the elevations (ft) for each placemark from the contour map to determine the **relief**. Find the slope of Mt. Saint Helens using the following formula (this value is representative for many composite cone volcanoes):
slope angle = tan^{-1} (relief ÷ horizontal distance).
- [ ] 5-10°
- [ ] 15-20°
- [ ] 25-35°
- [ ] 35-45°

(b) Check and double-click the placemarks for Problem 2b and turn on the "Mt. St. Helens Volcanic Features" map. Use the Ruler Tool [tool] to determine the greatest distance (in km) affected by the lateral blast of the May 18, 1980 eruption (measure between the placemarks for Problem 2b).
- [ ] 98-101 km
- [ ] 16-21 km
- [ ] 40-44 km
- [ ] 22-28 km

(c) Check and double-click placemark Problem 2c to fly to the town of Bronte near Mt. Etna, a large mountain on the island of Sicily. You are a geologic consultant for a company that is considering building a factory in Bronte. Based on your previous work on Mt. Saint Helens and on observations that you make as you fly over the region, decide which answer you would give the company.
- [ ] no, Bronte is about 15 km from an active composite cone volcano
- [ ] no, a previous lava flow has covered the eastern parts of town
- [ ] no, because of both of the previous reasons
- [ ] yes, the weather here is great

(d) Check and double-click the placemarks for Problem 2d to fly to Mt. Vesuvius in Italy and view the "Pompeii Pyroclastic Flow Flyover." Use the Ruler Tool [tool] to determine the distance (in km) from Mt. Vesuvius to the excavated ruins of the city of Pompeii (which was buried by volcanic ash and pyroclastic material from Mt. Vesuvius during the eruption of 79 C.E.) If a **nuée ardent** (pyroclastic surge) accompanies the next eruption of Mt. Vesuvius and it travels at 300 km/hr, how many <u>minutes</u> would it take to reach the site of ancient Pompeii?
- [ ] ~2 min
- [ ] ~0.2 min
- [ ] ~5 min
- [ ] ~7 min

(e) Check and double-click the placemark for Problem 2f to fly to Crater Lake, OR and turn on the "Crater Lake, Or map". The composite cone Mt. Mazama explosively erupted ~7000 years ago. Not only was most of the cone blown away, but a large depression was created as the material collapsed into the evacuated magma chamber. What is this depression called? *Note the parasitic* **cinder cone** *(Wizard Island) that has developed.*

☐ crater *(Crater Lake is named appropriately)*
☐ caldera *(Crater Lake is actually incorrectly named)*
☐ chamber basin *(Crater Lake is actually incorrectly named)*
☐ evacuation basin *(Crater Lake is actually incorrectly named)*

Cinder Cone Volcanoes

3. **Cinder cones** are the smallest volcanoes and often are found parasitic on larger volcanoes. They typically erupt pyroclastic materials, but they also commonly have small lava flows that emit from the base of the cone during the latter parts of their short-lived eruptive "lives".

(a) Check and double-click the placemarks for Problem 3a to fly to SP Mountain, AZ. Turn on the "SP Mountain contour map" in the Problem 3 folder. The brown lines are **contours** (lines of equal elevations). Use the Ruler Tool to determine the **horizontal distance** between placemarks for Problem 3a (*use feet as your units*) and then subtract the elevations (ft) for each placemark from the contour map to determine the **relief**. Find the slope of SP Mountain using the following formula (this value is representative for many cinder cone volcanoes): ***slope angle = tan^{-1} (relief ÷ horizontal distance).***

☐ 5-10°
☐ 15-20°
☐ 25-35°
☐ 35-45°

(b) Check and double-click the placemark for Problem 3b to fly to Menan Buttes, ID. These cinder cones are dominated by pyroclastic material and are very asymmetric. What causes the asymmetry (*think about the type of material and the direction of asymmetry*)?

☐ the cones have been reworked by the nearby river
☐ cinder cones are comprised of pyroclastic material, which can be blown by winds
☐ this is just a typical erosion pattern of cinder cones
☐ landsliding of material

(c) Double-click the placemarks for Problem 3c to fly to Sunset Crater, AZ. To get a sense of scale for cinder cones, use the Ruler Tool to determine Sunset Crater's width (in km).

☐ 10-15 km
☐ 5-10 km
☐ 15-20 km
☐ 1-3 km

To answer questions for this worksheet, go to the following Geotours folder in Google Earth:
2. Exploring Geology Using Google Earth > Geotour Worksheets > F. Sedimentary Rocks

F. Sedimentary

Rocks

Worksheet

Supplemental background material also may be available in your textbook, through various internet resources, and within files in the 2. Exploring Geology Using Google Earth > Geotour Site Library.

Sedimentary Rocks Exposed in the Grand Canyon
- Grand Canyon, AZ

Tilted and Folded Sedimentary Rocks
- Lewis Range, MT

Arid Depositional Environments
- Death Valley, CA

Fluvial Depositional Environments
- Niger Delta, Nigeria

35

Sedimentary Rocks

Sedimentary Rocks Exposed in the Grand Canyon

1. The Grand Canyon is a marvelous place to study geology. Here, weathering and erosion by the Colorado River have revealed an immense span of time as recorded in the rock record. Turn on the "Grand Canyon geologic map" in the **F. Sedimentary Rocks** worksheet folder and make it semi-transparent (*Note: This overlay is 4 mb and takes time to be downloaded into Google Earth's memory cache. Slight mismatches exist because the map is draped over topography.*) In this worksheet, we're going to focus on the **cover sequence** of sedimentary rocks that were deposited on top of the older metamorphic and igneous **basement** rocks.

(a) In order to get a sense of horizontal scale, check and double-click the placemarks for

Problem 1a. Use the Ruler Tool to determine the distance (in km) between the North and South Rims.
- [ ] 14-16 km
- [ ] 4-7 km
- [ ] 22-25 km
- [ ] 30-33 km

(b) To get a sense of relief in this area, check and double-click the placemarks for Problem 1b. Hover the hand cursor over each placemark and find the relief by subtracting the elevations shown at the bottom of the screen in the information bar (in m).
- [ ] 2310-2340 m
- [ ] 1400-1425 m
- [ ] 975-1000 m
- [ ] 3020-3050 m

(c) To identify the boundary between the sedimentary cover rocks (above) and the igneous and metamorphic basement rocks (below), check and double-click the placemarks for Problem 1c. These placemarks delineate the steep and narrow Inner Gorge adjacent to the Colorado River. Which of the following statements is <u>incorrect</u>?
- [ ] The crystalline metamorphic & igneous basement rocks are located in the Inner Gorge.
- [ ] The Inner Gorge is ~0.5-0.7 km wide here.
- [ ] The sedimentary cover rocks all erode to form uniformly similar slopes.
- [ ] The sedimentary cover rocks more easily erode to form the wider parts of the Canyon.

(d) Locate the lavender-colored Redwall Limestone (Mr) using the geologic map (placemark Problem 1d). What kind of slope does it form?
- [ ] steep cliff
- [ ] gentle slope

(e) Locate the white-colored Bright Angel Shale (Cba) using the geologic map (placemark Problem 1e). What kind of slope does it form?
- [ ] steep cliff
- [ ] gentle slope

(f) Locate the tan-colored Coconino Sandstone (Pc) using the geologic map (placemark Problem 1f). What kind of slope does it form?

☐ steep cliff

☐ gentle slope

(g) What general statement can you now make about the slopes of the Grand Canyon based on the different _sedimentary rock types_?

☐ all sedimentary rocks weather to form uniform slopes

☐ there is no consistent slope pattern related to type of sedimentary rock

☐ "strong" rocks (shales) form cliffs, "weak" rocks (limestones/sandstones) form slopes

☐ "strong" rocks (limestones/sandstones) form cliffs, "weak" rocks (shales) form slopes

(h) Check and double-click the placemarks for Problem 1h (_you also may want to look at the "Rock Units" folder in the Geotour Site Library "Grand Canyon, AZ" folder for this chapter in addition to the geologic map_). Note the following sequence of Cambrian rock units: Tapeats Sandstone (near shore; lowest rock unit), Bright Angel Shale (intermediate distance from shore; middle rock unit), and Muav Limestone (distal from shore sediments, but still shallow water; top rock unit). The vertical stacking of these units represents either a **regression** (fall) or **transgression** (rise) of sea level. Which is occurring in this area?

☐ regression (fall in sea level)

☐ transgression (rise in sea level)

(i) Check and double-click Problem 1i. Here, the Colorado River narrows at the mouth of a side canyon, and there are some rapids. What causes this?

☐ a delta containing sand and gravel from a side canyon

☐ a point bar of sand and gravel at the inside bend of a meandering river

☐ fine silts from a river floodplain

Tilted and Folded Sedimentary Rocks

2. As we discovered from Problem 1 in the Grand Canyon, one of the most easily observable sedimentary structures is **bedding** or **layering**. However, not all regions have experienced the spectacular erosion that the Grand Canyon has. In mountainous areas (such as the Lewis Range in Montana), plate tectonic convergence has exposed various sedimentary layers by tilting, folding, and/or faulting them. Resistant rocks still erode to form cliffs, and less resistant rocks still form low slopes or valleys. Folding and faulting can repeat the same sedimentary layers multiple times, creating a series of repeating ridges (resistant) and valleys (less resistant).

(a) Check and double-click the placemarks for Problems 2a-i and -ii. Which of the following is true regarding the resistance to erosion of the layers specified by the placemarks?

☐ Problem 2a-i is less resistant and Problem 2a-ii is more resistant

☐ Problem 2a-i is more resistant and Problem 2a-ii is less resistant

☐ Problem 2a-i and Problem 2a-ii have similar resistances to erosion

(b) Based on the geometry of these layers, which direction are the rocks tilted (geoscientists might ask which way do the rock layers "**dip**")? _Hint: Tilted sedimentary rocks typically erode to form an asymmetric ridge called a_ **flatiron**—_the dip slope of such a ridge lies parallel to bedding, forming a somewhat planar surface with a pointed or rounded edge pointing in the direction opposite of the tilt (geoscientists thought that this geometry resembled the flat bottom of a laundry iron). Put another way: if you poured water on the flat dip slope, in which direction would the water flow?_ Problem 2b is a polygon outlining one of the flatirons.

☐ The layers dip west

☐ The layers dip east

37

(c) Now that you've seen some flatirons, check and double-click the placemarks for Problems 2c-i and -ii. Which choice best describes the dip/tilt of the layers for each placemark?
- [ ] both placemarks are tilted east
- [ ] both placemarks are tilted west
- [ ] Problem 2c-i is tilted west and Problem 2c-ii is tilted east
- [ ] Problem 2c-i is tilted east and Problem 2c-ii is tilted west

---

Arid Depositional Environments

3. Check and double-click placemark Problem 3a to fly to Death Valley, CA.
(a) What depositional environment does placemark Problem 3a represent?
- [ ] sand dunes *(forming sandstone)*
- [ ] alluvial fan *(forming arkose & conglomerate)*
- [ ] playa lake *(forming evaporite rocks)*
- [ ] mountain stream *(forming large boulder/cobble conglomerates)*

(b) What depositional environment does placemark Problem 3b represent?
- [ ] sand dunes *(forming sandstone)*
- [ ] alluvial fan *(forming arkose & conglomerate)*
- [ ] playa lake *(forming evaporite rocks)*
- [ ] mountain stream *(forming large boulder/cobble conglomerates)*

Fluvial Depositional Environments

4. Check and double-click the placemarks for Problems 4a-i and 4a-ii to fly to the Niger Delta, Nigeria. The Niger Delta consists of sediment carried to the sea by the Niger River. Check the "What a Geologist Sees" folder in the Problem 4 folder to activate a Niger Delta overlay that provides the approximate present-day outline of the delta with features created by running water individually labeled.

(a) What type of sedimentary rock will likely form at each of the placemarks in their present day setting?
- [ ] 4a-i: sandstone/conglomerate and 4a-ii: shale/siltstone/coal
- [ ] 4a-i: shale/siltstone/coal and 4a-ii: sandstone/conglomerate

(b) If you owned the plant at placemark Problem 4b, which of the following concerns would you have?
- [ ] the meander would continue migrating towards my buildings
- [ ] the meander would continue migrating away from my buildings

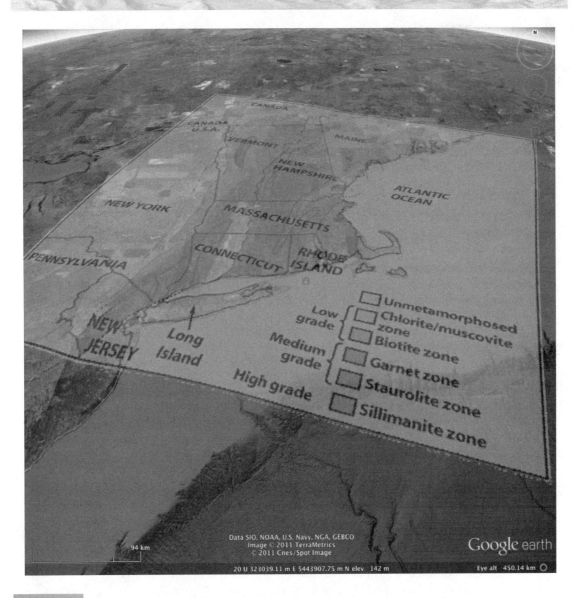

Metamorphic Rocks

To answer questions for this worksheet, go to the following Geotours folder in Google Earth:
2. Exploring Geology Using Google Earth > Geotour Worksheets > G. Metamorphic Rocks

Supplemental background material also may be available in your textbook, through various internet resources, and within files in the 2. Exploring Geology Using Google Earth > Geotour Site Library.

G. Metamorphic

Rocks

Worksheet

Zones of Metamorphic Grade
- New England, USA

Types of Metamorphism
- New England, USA
- Henry Mountains, UT
- Baraboo, WI

Metamorphism: Past & Present
- Global Map of Shields & Mt. Belts
- Canadian Shield, Hudson Bay, Canada

Geotour Worksheet G
Metamorphic Rocks

Zones of Metamorphic Grade

1. In the Problem 1 folder within the **G. Metamorphic Rocks** worksheet folder, turn on the "New England Metamorphic Zones" map overlay (double-click it to fly to the area).
(a) If you traveled north from placemark Problem 1a, would you go into higher or lower **metamorphic grade** rocks?
   ☐ lower metamorphic grade
   ☐ higher metamorphic grade
(b) Based on your answer to (a), would the **metamorphic core** (the zone of highest metamorphism) of this region be near placemark Problem 1b-i or 1b-ii?
   ☐ 1b-i
   ☐ 1b-ii
(c) For the schist pictured in placemark Problem 1c, which zone did it likely occur in (*identify the red minerals*)?
   ☐ unmetamorphosed
   ☐ chlorite/muscovite zone
   ☐ biotite zone
   ☐ garnet zone

Types of Metamorphism

2. For this problem, you will compare and contrast the metamorphic area depicted in Problem 1 with an area in the Henry Mountains of Utah.
(a) Referring to the area covered by the "New England Metamorphic Zones" map overlay in Problem 1, what type of metamorphism likely produced these metamorphic rocks?
   ☐ shock metamorphism (*e.g., impact of a meteorite*)
   ☐ contact metamorphism (*e.g., intrusion of igneous sills & dikes*)
   ☐ regional metamorphism (*e.g., burial and/or large-scale convergent tectonics*)
(b) Check and double-click placemark Problem 2b to fly to the Henry Mountains in Utah. What type of metamorphism likely produced the thin, white bleaching of the reddish sandstone in this area? *Hint: You may want to fly around this area to look at the features present.*
   ☐ shock metamorphism (*e.g., impact of a meteorite*)
   ☐ contact metamorphism (*e.g., intrusion of igneous sills & dikes*)
   ☐ regional metamorphism (*e.g., burial and/or large-scale convergent tectonics*)
(c) Check and double-click the Van Hise Rock, WI photo overlay in the Problem 2 folder to fly to the Rock Springs, WI area to view a zoomable photo of the outcrop. These Precambrian rocks experienced greenschist grade metamorphism as the rock layers were buried and then subsequently folded and sheared past each other. Which of the choices below best describes the type of metamorphism that these rocks experienced?
   ☐ shock metamorphism (*e.g., impact of a meteorite*)
   ☐ contact metamorphism (*e.g., intrusion of igneous sills & dikes*)
   ☐ regional metamorphism (*e.g., burial and/or large-scale convergent tectonics*)

(d) The **protolith** (parent rock) of the pink/maroon rock at Van Hise Rock was a pure quartz sandstone, whereas the darker layer was originally a lens of clay-rich shale. What metamorphic rocks did these protoliths (sandstone, shale) become?

☐ quartzite, marble
☐ quartzite, phyllite
☐ slate, marble
☐ marble, gneiss

Metamorphism: Past & Present

3. Check the box next to "Global Map of Shields & Mt Belts" in the Problem 3 folder to view the Precambrian shields (maroon) and the younger mountain belts (green). Also, in the Layers panel, turn on *Gallery > Earthquakes* and *Gallery > Volcanoes*.

(a) Fly around the world and zoom in until you see the earthquake and volcano placemarks. Using these as a proxy for tectonic activity, which of the statements below best describes your observations?

☐ Precambrian shields are less tectonically active than younger mountain belts
☐ Precambrian shields are more tectonically active than younger mountain belts
☐ Precambrian shields and younger mountain belts are similarly tectonically active

(b) Turn off the "Global Map of Shields & Mt Belts" overlay. Check and double-click placemark Problem 3b to fly to the Canadian Shield near Hudson Bay. Here, metamorphic foliations have differentially eroded to show two distinct foliation trends of different ages (one truncating the other). What evidence do you see in the foliations to suggest that these rocks were once deeply buried and experienced tectonic deformation?

☐ certain rock layers have developed basins which now contain water
☐ there is a distinct cleavage in the rocks when viewed closely
☐ the foliations are deformed into contorted folds like we see in many gneisses
☐ these rocks show no evidence of tectonic deformation

Earthquakes

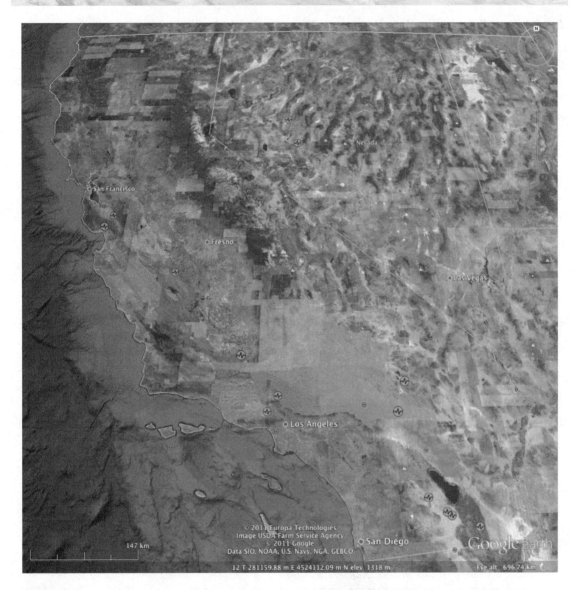

To answer questions for this worksheet, go to the following Geotours folder in Google Earth:
2. Exploring Geology Using Google Earth > Geotour Worksheets > H. Earthquakes

Supplemental background material also may be available in your textbook, through various internet resources, and within files in the 2. Exploring Geology Using Google Earth > Geotour Site Library.

H. Earthquakes

Worksheet

Evidence of Earthquake Activity along a Plate Boundary
- San Andreas Fault, CA

Evidence of Intraplate Earthquake Activity
- New Madrid, MO

Stress Transfer and Earthquake Prediction
- North Anatolian Fault, Turkey

Tsunami Devastation
- Banda Aceh, Sumatra

Earthquakes

Evidence of Earthquake Activity along a Plate Boundary

1. Check and double-click the placemarks for Problem 1a in the **H. Earthquakes** worksheet folder to fly to the Wallace Creek area along the San Andreas fault in California. Also, in the Layers panel, turn on *Gallery > Earthquakes*.

(a) Use the Ruler Tool [image] to determine the distance (in m) that Wallace Creek has been offset by recent motion along the San Andreas (measure between placemarks 1a).
- [ ] ~75 m
- [ ] ~100 m
- [ ] ~303 m
- [ ] ~160 m

(b) Assume that the present rate of slip along the fault is about 6 cm/yr. How many years has it taken to develop the observed offset of the stream?
- [ ] ~2667 yrs
- [ ] ~5033 yrs
- [ ] ~1010 yrs
- [ ] ~10,100 yrs

(c) What is the sense of offset along the fault? (i.e., Walk along either segment of Wallace Creek toward the fault. At the fault, determine the direction you turn to find the offset segment of the stream across the fault.)
- [ ] right lateral
- [ ] left lateral

(d) Use the Ruler Tool [image] to determine the heading/orientation (in degrees) of this segment of the San Andreas fault. Note that orientations of linear features have two directions on a compass (e.g., 000°<->180°=N/S, 045°<->225°=NE/SW, etc.).
- [ ] 050° <-> 230° *(fault oriented NE/SW)*
- [ ] 000° <-> 180° *(fault oriented N/S)*
- [ ] 318° <-> 138° *(fault oriented NW/SE)*
- [ ] 283° <-> 103° *(fault oriented WNW/ESE)*

(e) Check and double-click Problem 1e to fly to the Palmdale area of the San Andreas fault. What type of feature along the fault trace does the placemark point to?
- [ ] offset stream *(stream whose course has been changed due to slip on the fault)*
- [ ] sag pond *(a depression along the fault trace filled with water)*
- [ ] fault scarp *(surface exposure of the fault plane)*
- [ ] fault gouge *(ground up rock due to slip on the fault)*

(f) Check and double-click the placemarks for Problem 1f to fly to another offset stream along the San Andreas. Use the Ruler Tool [icon] to determine the distance (in m) that the stream has been offset and then use the present rate of slip along the fault (6 cm/yr) to determine how many years it would take to cause this much offset along the fault.

- ☐ 20,000-25,000 yrs
- ☐ 8000-10,000 yrs
- ☐ 4000-5000 yrs
- ☐ 800-1000 yrs

(g) Traces of the San Andreas fault system pass through the San Francisco/Oakland area and then skirt along the shoreline for some distance to the northwest. Check and double-click Problem 1g to fly to the coastal area near Fort Ross. Which of the statements is <u>incorrect</u>?

- ☐ the San Andreas fault created NE/SW-oriented valleys
- ☐ rocks on the west side of the San Andreas fault move to the NW

Just for Fun..._Consider your answers regarding the length of time it has taken to create the two offset streams...recall that the answers were very different! Can you think of hypotheses that might explain this discrepancy?_

Evidence of Intraplate Earthquake Activity

2. Turn on the "New Madrid, Mo Seismicity" map overlay and double-click it to fly to the Missouri, Illinois, Kentucky, Arkansas, and Tennessee area. Also, in the Layers panel, turn on _Gallery > Earthquakes_. This area is well within the North American plate, yet it has a surprising number of earthquakes (you may recall that this area was the site of three magnitude 8.0-8.5 earthquakes during the winter of 1811-1812).

(a) Using the map overlay data, what is the dominant trend/pattern of earthquakes in this region (i.e., what direction would you assume that most of the faults are oriented)?

- ☐ E-W
- ☐ N-S
- ☐ NE-SW

(b) What is the best explanation for this <u>intraplate</u> seismicity?

- ☐ this is actually a plate boundary that is covered by Mississippi River sediments
- ☐ reactivation of ancient faults underlying the region due to loading and/or distant stresses
- ☐ upwelling magma from a hot spot similar to earthquakes at Yellowstone National Park
- ☐ subsidence and compaction of Mississippi River deposits

(c) On the overlay, there is a concentration of higher magnitude earthquakes (blue dots) near New Madrid, MO (location of the early 19th century large earthquakes). Which statement is the best hypothesis to explain this observation?

- ☐ there may be a bend in the underlying fault systems
- ☐ there may be an intersection of several faults at this location
- ☐ older earthquakes may have slipped elsewhere, placing more stress at this location
- ☐ all of these hypotheses are viable and worth testing

Stress Transfer and Earthquake Prediction

3. Check and double-click the "North Anatolian Fault, Turkey" placemark in the Problem 3 folder to fly to northern Turkey. Also, in the Layers panel, turn on *Gallery > Earthquakes*. The figure within the placemark shows a historical summary of earthquake activity along the North Anatolian fault (including the segments of the fault that were active during each event).

(a) What is the dominant pattern of major earthquakes along the North Anatolian fault from **1939-1967**?

☐ generally become progressively younger from west (oldest) to east (youngest)

☐ generally become progressively younger from east (oldest) to west (youngest)

☐ There is no apparent pattern

(b) If a pattern existed during that time, what might cause this pattern?

☐ slip from the rupture of one earthquake increased stress on adjacent fault segments

☐ slip from the rupture of one earthquake decreased stress on adjacent fault segments

☐ There is no apparent pattern, so the earthquakes occured randomly along the fault

(c) In the lower figure of the placemark, look at the segments of the North Anatolian fault that have slipped (ruptured, white) versus those that have not ruptured recently (black). Where would you predict future earthquakes along this fault?

☐ Ankara area

☐ Erzincan area

☐ Yedisu area

Tsunami Devastation

4. Check and double-click the "Regional Plate Tectonics" map overlay and the Epicenter placemark in the Problem 4 folder to fly to the Sumatra region that was devastated by a tsunami in late 2004. Also, in the Layers panel, turn on *Gallery > Earthquakes* and *Gallery > Volcanoes*.

(a) Which tectonic plate was subducted beneath which overriding tectonic plate at this locality to produce this massive earthquake? *Note that the subduction zone is delineated by the line with triangles and that the volcanoes will be on the overriding plate.*

☐ Indian Plate is being subducted beneath the Burma Plate

☐ Burma Plate is being subducted beneath the Indian Plate

☐ Sunda Plate is being subducted beneath the Burma Plate

☐ Burma Plate is being subducted beneath the Sunda Plate

(b) Check and double-click the Tsunami Travel-Time Animation placemark in the Problem 4 folder. How long did it take the tsunami to first arrive at Sri Lanka across the Indian Ocean?

☐ 50 minutes

☐ 90 minutes

☐ 120 minutes

☐ 160 minutes

(c) Use the Ruler Tool [icon] to determine the distance from the earthquake epicenter to placemark Problem 4c on Sri Lanka in (km). Use your answer from Problem 4b to determine the average velocity at which the tsunami traveled across the Indian Ocean (km/hr).

☐ 150-225 km/hr

☐ 750-825 km/hr

☐ 350-425 km/hr

☐ 1250-1325 km/hr

Geologic Structures

To answer questions for this worksheet, go to the following Geotours folder in Google Earth:
2. Exploring Geology Using Google Earth > Geotour Worksheets > I. Geologic Structures

Supplemental background material also may be available in your textbook, through various internet resources, and within files in the 2. Exploring Geology Using Google Earth > Geotour Site Library.

I. Geologic Structures Worksheet

Calculating Strike and Dip from Flatirons
- Wind River Mountains, WY

Brittle Structures-Joints
- Arches National Park, UT

Brittle Structures-Faults
- Canyonlands National Park, UT
- Squaw Point, OR
- Canadian Front Ranges, Canada
- San Andreas Fault, CA

Ductile Structures-Folds
- Zagros Fold Belt, Iran
- Sheep Mountain, WY
- Milton, PA
- Quail Creek State Park, Hurricane, UT

Geotour Worksheet I
Geologic Structures

Calculating Strike and Dip from Flatirons

1. Check and double-click the polygon "Flatiron" in the Problem 1 folder within the **I. Geologic Structures** worksheet folder to fly to the northeastern flank of the Wind River Mountain range in Wyoming. You may recall from the sedimentary rock worksheet that tilted sedimentary rocks typically erode to form asymmetric ridges called **flatirons**—a relatively planar surface with a pointed or rounded edge pointing in the direction opposite of the tilt (*geoscientists thought that this geometry resembled the flat bottom of a laundry iron, hence the name*). In this view, the polygon outlines one of the many flatirons that border the uplift (flatirons are common in folded/tilted sedimentary strata). To make it easier to see, you might consider temporarily increasing the vertical exaggeration to 2 in *Google Earth*.

(a) Turn on the Problem 1a-strike placemark. You probably recall from your text that geoscientists describe the orientation of structures by measuring the **strike** and **dip** of the rock layers. The **strike** is the angle between an imaginary horizontal line on the structure and the direction of true north. To determine an approximate strike, use the Ruler Tool (Line tab, use whatever units the elevation in the Info Bar at the bottom of the screen is using, preferably meters) to determine the elevation at the point of the placemark icon by hovering the crosshairs over it and looking at the Info Bar (somewhere between 1830-1840 m...let's use 1835 m). Now draw a line from the Problem 1a-strike placemark using the Ruler tool to the right such that you click the second point of the line at a location at the exact same elevation (~1835 m). This is the horizontal strike line. What is the orientation (000°-360°)?

☐ 230-240°
☐ 040-050°
☐ 310-320°
☐ 125-135°

Note: This approximation works because we're assuming that the flat part of the flatiron approximates a bedding plane surface.

(b) Leave the 1835 m line on the screen and turn on the Problem 1b-dip placemark. Recall that **dip** is the angle of a structure's slope measured in a vertical plane perpendicular to the strike. To determine this perpendicular dip direction, draw a perpendicular line <u>from the 1835 m line to the placemark point</u> (use the Ruler tool; the strike line will disappear and you will now see only the dip line). The orientation of the dip line is the **dip direction** (the direction that water would flow down the flatiron). What is the dip direction (000°-360°)?

☐ 040-050°
☐ 170-180°
☐ 210-220°
☐ 310-320°

(c) Note the length of the dip line in meters (or the same units that your elevations are in) and hover over the point of placemark Problem 1b-dip to determine its elevation. What is the approximate dip angle (00°-90°)? To determine the amount of dip, use the following formula: **dip = tan⁻¹ ([elevation difference between placemark Prob 1b-dip & strike line] ÷ length of the dip line).**

☐ 05-12°
☐ 19-26°
☐ 31-37°
☐ 43-48°

Just for Fun..._Determining the strike and dip of a flatiron is harder to explain in words than it is to actually do in Google Earth! Find some other nearby flatirons (look for the most planar ones) and try to draw a horizontal strike line. How does it compare to your answer? Do the same for dip and see how close you come. Think about reasons why you might have differences._

Brittle Structures-Joints

2. In the Problem 2 folder, check and double-click placemark Problem 2 to fly to Arches National Park in Utah. Also, turn on the "Arches NP Geologic Map-North" map overlay and make it semi-transparent. Placemark Problem 2 points to a road (dashed red line on the geologic map) that approximates the **axis of a salt-filled anticline** that has collapsed to form Salt Valley. Along the flanks of this anticlinal valley, some of the rock units exhibit a prominent linear pattern that reflects systematic **joint sets** that have developed in association with the Salt Valley anticline and subsequent collapse.

(a) The Problem 2a placemark points to the youngest rock unit that can be easily observed to contain these joint sets. Using the geologic map overlay, which unit is this? _Note: The oldest rock unit in the map key is the Paradox Formation and the youngest is the Modern Alluvial Deposits (older rock units are listed on the bottom and younger ones are on top)._

☐ Jcd
☐ Jes
☐ Jctm
☐ Jmt

(b) What happens to the joint sets beneath the younger rock units (Jmt & Jms) at placemark Problem 2b?

☐ they do not exist beneath placemark Problem 2b anywhere in the subsurface
☐ they continue on beneath placemark Problem 2b in the white unit (Jctm) because the younger Jmt/Jms units just cover them

Just for Fun..._Think about your answer for 2b...does this help date the age of jointing or does it just reflect how joints develop differently in different rock types?_

(c) Check and double-click the placemark for Problem 2c, which flies you to an area on the NE side of Salt Valley that shows two prominent joint sets. Joint sets may intersect at a variety of angles, but most commonly are classified as being at right angles (**orthogonal**) or not at right angles. Which kind are these?

☐ right angles (orthogonal)
☐ not at right angles

(d) Arches National Park is known for its spectacular **arches** that develop in the jointed areas of the park. Placemark Problem 2d flies you to Landscape Arch, the arch with the largest span in Arches National Park.   If the stresses that stretched the rock to form the joints were oriented <u>perpendicular</u> to the joints (and the rock fins that contain the arches), in what direction were the stresses oriented?
☐ N-S
☐ E-W
☐ NW-SE
☐ NE-SW

Brittle Structures-Faults

3. Faults are typically classified in one of three main categories: **normal, reverse,** and **strike-slip**. The following questions will take you to locales that provide examples of each type of these faults.

(a) Check and double-click the placemarks for Problems 3a-i, -ii, and -iii to fly to Canyonlands National Park-Needles District in Utah. This area is a region in the Grabens area that is experiencing **normal faulting** as the rock units in the region are extending to the west. It is called the Grabens because the faulting forms fault-bounded blocks that have been down-dropped (**grabens**) with intervening high blocks (**horsts**). Which of the following is the best answer for identifying the structures in the view?
☐ 3a-i: graben
☐ 3a-ii: horst
☐ 3a-i and 3a-iii: grabens
☐ 3a-i and 3a-iii: horsts

(b) Check and double-click placemark Problem 3b to fly to Squaw Point, OR. The placemark lies on a normal **fault scarp**, an offset of the land surface by slip along a fault. Which side of the normal fault is the down-dropped hanging wall?
☐ left (SW)
☐ right (NE)

(c) Placemark Problem 3c highlights another area east of Problem 3b that has experienced extension along a series of normal faults, forming a series of **half grabens**. If the extensional stresses that are stretching the rocks are approximately perpendicular to the fault traces, which direction are the stresses oriented?
☐ N-S
☐ E-W
☐ NW-SE
☐ NE-SW

(d) Photo Problem 3d is almost a cross-sectional view that shows a low-angle **reverse fault** (the McConnell **thrust fault**) placing older Paleozoic rocks (gray) on top of forested Mesozoic rocks. The gray rocks are interpreted to have been uplifted over 5 km and transported over 40 km from their original site of deposition. Which direction were the rocks transported? *You might also turn on the "Kananaskis Country Geologic Map, Canada" map overlay and adjust the photo's transparency (triangles are on the hanging-wall side of the fault).*
☐ to the right (east)
☐ to the left (west)

(e) Check and double-click the placemarks for Problem 3e, and turn on the "Kananaskis Country Geologic Map, Canada" overlay. Reverse faults, and particularly thrust faults, duplicate rock units by stacking the same units on top of each other over and over. How many <u>major</u> repetitions of the Devonian Palliser Formation (DPa, cyan colored) are seen in the rocks north of the Bow River? *Count the repetitions marked by the placemarks.*

☐ 0
☐ 1
☐ 3
☐ 9

(f) Photo Problem 3f is almost a cross-sectional view that shows a zoomable photo of a reverse fault near Barrier Lake, Canada (*turn on the "Kananaskis Country Geologic Map, Canada" overlay and adjust the photo's transparency*). Which direction were the rocks above the fault displaced?

☐ to the right (east)
☐ to the left (west)

(g) Check and double-click placemark Problem 3g to fly to the San Andreas fault zone near the Carrizo Plain in California. This transform plate boundary is also a **strike-slip fault**. Given the offset stream in the view, what is the sense of offset along this fault?

☐ left-lateral
☐ right-lateral

Ductile Structures-Folds

4. Folds are classified by a wide range of names, depending on their geometry: **anticlines, synclines, domes, basins**, and **monoclines**. The following questions will take you to locales that provide examples of some of these remarkable structures.

(a) Check and double-click placemark Problem 4a to fly to the Zagros Fold Belt in Iran. What type of fold is this? *Hint: Look at the **flatirons** on the fold limbs.*

☐ anticline *(beds generally dip away from center **fold hinge** with oldest rocks at the center)*
☐ syncline *(beds generally dip towards center **fold hinge** with youngest rocks at the center)*

(b) Check and double-click the placemarks for Problems 4b-i and 4b-ii. Folds are said to be **upright** in areas where their fold hinge is approximately horizontal (<u>bedding contacts tend to be straight and parallel to the main fold hinge</u>) and **plunging** in areas where their fold hinge is inclined into the ground (<u>bedding contacts tend to curve and wrap around from one **fold limb** to another</u>). Which of the following is correct?

☐ 4b-i: plunging and 4b-ii: plunging
☐ 4b-i: upright and 4b-ii: upright
☐ 4b-i: plunging and 4b-ii: upright
☐ 4b-i: upright and 4b-ii: plunging

(c) For placemark Problem 4c, which direction is the fold plunging? *Anticlines plunge in the direction that the rock layers are convex as they wrap around from one fold limb to another, whereas synclines plunge in the direction that the rock layers are concave as they wrap around from one fold limb to another. Commonly, adjacent anticlines and synclines plunge in the same direction. To visualize this, fold a piece of paper into an anticline/syncline and plunge it into a table to see the map pattern (trace the intersection of the paper and the table).*

☐ NW
☐ SE

(d) Check and double-click placemark Problem 4d to fly to Sheep Mountain in Wyoming. Turn on the "Sheep Mt. Geologic Map" overlay. The markers with numbers show the orientation of bedding—the longer line shows the **strike direction** and the shorter line shows the **dip direction**. The number indicates the **angle of dip** (e.g., the NE fold limb has dips to the NE ranging from 40's to the 80's, whereas the SW fold limb has dips to the SW from the high 20's to the 40's). The types of fold axes and the colored rock units (oldest listed at the bottom [Mm] and becoming progressively younger towards the top [Kmr]) are identified in the map legend. Which of the following best describes the area around placemark Problem 4d?

☐ asymmetric anticline plunging NW
☐ asymmetric syncline plunging NW
☐ asymmetric anticline plunging SE
☐ asymmetric syncline plunging SE

(e) Using the fold axis and the map pattern of the rock units at placemark Problem 4e, what is this type of fold and which direction is it plunging?

☐ NW-plunging anticline
☐ NW-plunging syncline
☐ SE-plunging anticline
☐ SE-plunging syncline

(f) Which direction is the fold at placemark Problem 4f plunging and what type of fold is it? *Oftentimes the plunging "noses" of anticlines are long and tapering, whereas the plunging "noses" of synclines are more blunt and rounded.*

☐ WSW-plunging anticline
☐ WSW-plunging syncline
☐ ENE-plunging anticline
☐ ENE-plunging syncline

(g) Placemark Problem 4g points to a prominent flatiron of Triassic Shinarump Conglomerate on a fold limb. What name best describes the overall fold? *You might turn on the "Geologic Map of Quail Creek SP, UT" map overlay and make it semi-transparent to assist your interpretation (check out the cross section in the map legend as well!).*

☐ SW-plunging anticline
☐ SW-plunging syncline
☐ NE-plunging anticline
☐ NE-plunging syncline

(h) Which direction do the beds at placemark Problem 4h <u>strike</u>?

☐ NE
☐ SE
☐ SW
☐ NW

(i) Which direction do the beds at placemark Problem 4i <u>dip</u>?

☐ NE
☐ SE
☐ SW
☐ NW

Geotour Worksheet J
Geologic Time

J. Geologic

Time

Worksheet

To answer questions for this worksheet, go to the following Geotours folder in Google Earth:
2. Exploring Geology Using Google Earth > Geotour Worksheets > J. Geologic Time

Supplemental background material also may be available in your textbook, through various internet resources, and within files in the 2. Exploring Geology Using Google Earth > Geotour Site Library.

Relative Age Dating & Unconformities
- Grand Canyon, AZ
Stratigraphic Formations in Southern Utah
- Grand Staircase, UT
- Zion National Park, UT

Rock Layers and Monoclines, Circle Cliffs, UT
- Circle Cliffs, UT

Geotour Worksheet **J**
Geologic Time

Relative Age Dating & Unconformities

1. Check and double-click the placemarks for Problems 1a-i, -ii, -iii, and -iv in the **J. Geologic Time** worksheet folder to fly to the east end of the Grand Canyon.

(a) According to the *principle of original continuity*, which layer corresponds to the layer indicated by placemark Problem 1a-i?
- ☐ Problem 1a-ii
- ☐ Problem 1a-iii
- ☐ Problem 1a-iv

(b) According to the *principle of superposition*, what is the order of age of layering for placemarks Problem 1b-i (pink), -ii (dark), and -iii (variegated) from oldest to youngest?
- ☐ 1b-i, 1b-ii, 1b-iii
- ☐ 1b-iii, 1b-ii, 1b-i
- ☐ 1b-ii, 1b-iii, 1b-i
- ☐ 1b-i, 1b-iii, 1b-ii

(c) According to the *principle of cross-cutting relationships*, which layer is younger: placemark Problem 1c-i (pink) or 1c-ii (tan)? *Hint: trace the pink layer to the left and see what happens to it.*
- ☐ 1c-i
- ☐ 1c-ii

(d) The surface dividing the units in Problem 1c is called the "Great Unconformity" because it separates older layered Precambrian units from the younger sedimentary Cambrian unit with up to a 1 billion year gap in the rock record due to erosion. What type of unconformity is it at this location?
- ☐ nonconformity
- ☐ disconformity
- ☐ angular unconformity

(e) Check and double-click placemark Problem 1e to fly west to another exposure of the Great Unconformity. Here metamorphic rock and an igneous dike are truncated and overlain by the same tan sedimentary Cambrian unit. Which is older: the dike or the tan rock layer?
- ☐ dike
- ☐ tan sedimentary Cambrian rock layer

(f) What type of unconformity is the Great Unconformity at this location?
- ☐ nonconformity
- ☐ disconformity
- ☐ angular unconformity

(g) According to the *principle of inclusions*, which rock unit should have inclusions of the other?
- ☐ the dike & metamorphic rock should have inclusions of the tan Cambrian rock layer
- ☐ the tan sedimentary Cambrian rock layer should have inclusions of the dike & metamorphic rock

(h) The placemark labeled Problem 1h points to another unconformity between the Early Mississippian Redwall Limestone (below; steep cliff) and the Early Pennsylvanian Supai Group (above). The unconformity represents a hiatus of about 25 million years. What type of unconformity is this? *Hint: You know that both units are sedimentary. From the view you have, determine if the layers above and below the unconformity are parallel.*

☐ nonconformity

☐ disconformity

☐ angular unconformity

Just for Fun...*Turn on the "Grand Canyon Geologic Map" overlay and explore the region in Google Earth (perhaps first look at the "Stratigraphic Overview" placemark). Not only can you find multiple examples of the principles used to determine relative dating, but the map also shows you what geologic periods these units have been correlated with (which have been dated using absolute age-dating techniques). Note that the map is 4 mb in size and will take a moment to load.*

Stratigraphic Formations in Southern Utah

2. The Grand Canyon provides spectacular exposures of Paleozoic and Precambrian rock layers. As you travel north from the Grand Canyon, you encounter laterally extensive younger Mesozoic rock layers that have eroded into a series of cliffs and benches. Because of this stair-like erosional profile caused by differences in rock type, the area is known as the "Grand Staircase". The Moenave Formation forms most of the "Vermillion Cliffs", the Navajo Sandstone forms the "White Cliffs", and the Claron Formation forms the "Pink Cliffs".

(a) Turn on the "Geologic Map of Grand Staircase, UT" map overlay and make it semi-transparent. Check and double-click the placemarks for Problems 2a-i, -ii, and -iii. Which placemark corresponds to the correct cliff-forming unit? Use the "Map Symbols" section of the map to assist you. *You might want to temporarily change the vertical exaggeration to "3" to see the "steps".*

☐ 2a-i: Moenave; 2a-ii: Navajo; and 2a-iii: Claron

☐ 2a-i: Claron; 2a-ii: Navajo; and 2a-iii: Moenave

☐ 2a-i: Navajo; 2a-ii: Claron; and 2a-iii: Moenave

☐ 2a-i: Moenave; 2a-ii: Claron; and 2a-iii: Navajo

(b) The steep cliffs of Zion Canyon at Zion National Park in Utah comprise one of the "steps" in the Grand Staircase. Check and double-click placemark Problem 2b and turn on the "Geologic Map of Zion NP" to determine which cliff forms Zion Canyon.

☐ White Cliffs

☐ Vermillion Cliffs

☐ Pink Cliffs

(c) Checkerboard Mesa, one of the most famous landmarks in Zion National Park, is comprised of this same cliff-forming unit. Check and double-click the Checkerboard Mesa folder to fly there. The "checkerboard" pattern is due to the intersection of sandstone cross beds (yellow) with joints (green). Which of these formed first (*one of them formed while the rock was sediment*)?

☐ joints

☐ cross-beds

Rock Layers and Monoclines, Circle Cliffs, UT

3. The Circle Cliffs area allows us to not only look at the ages of some of the different rock layers exposed in the Colorado Plateau, but to also explore a geologic structure that is commonly associated with the Colorado Plateau area: the **monocline**. The Circle Cliffs erode into the monocline that forms the Waterpocket Fold of Capitol Reef National Park in Utah.

(a) Turn on the "Geologic Map of Circle Cliffs, UT" map overlay and locate the flatirons of Jurassic Kayenta (Jk) and Wingate (Jw) on the west and east sides of the Circle Cliffs area (placemarks 3a-i and -ii are two representative flatirons on the different limbs of the fold). Monoclines typically have a steeply tilted fold limb (**forelimb**) and then a shallowly dipping (sometimes sub-horizontal) **backlimb**. Which placemark corresponds to the steep forelimb? *You may want to temporarily set the vertical exaggeration to "3".*

☐ 3a-i

☐ 3a-ii

(b) Both monoclines and anticlines uplift and deform rock layers such that when eroded, certain ages of rocks are exposed in the center of the structure relative to the flanking limbs. What is the relative ages of the rocks exposed in the Circle Cliffs area (placemark Problem 3b)? *You may want to use the Map Symbols legend where older rocks are listed at the bottom (Permian formations) with progressively younger units above (Uncons. Quat. deposits).*

☐ older rocks are exposed in the center and younger rocks in the flanking flatirons

☐ younger rocks are exposed in the center and older rocks in the flanking flatirons

Earth History

K. Earth	To answer questions for this worksheet, go to the following Geotours folder in Google Earth:
	2. Exploring Geology Using Google Earth > Geotour Worksheets > K. Earth History
History	*Supplemental background material also may be available in your textbook, through various internet resources, and within files in the 2. Exploring Geology Using Google Earth > Geotour Site Library.*
Worksheet	

Paleogeography of the Earth
  • Paleogeographic maps of the Earth

Paleogeography of the Earth

1. Check the "Paleogeographic Maps" folder in the "Global Paleogeographic Model" folder within the **K. Earth History** worksheet folder and make the entire folder semi-transparent. Turn on the latitude and longitude lines (*View > Grid*) and *Borders and Labels* in the Layers panel. Now, play the time animation through several times and rotate the globe to watch the animation from different perspectives. You can manually play the animation by grabbing the right part of the time slider and moving it, or you can toggle the animation on/off automatically by clicking the right-most clock icon in the historical animation box (upper left corner of the *Google Earth* viewer). *Leave the Paleogeographic Maps folder checked to answer the following questions.*

(a) Check and double-click placemark Problem 1a. Approximately 600 Ma (make sure that the time slider in the upper left corner of the *Google Earth* viewer window is all the way to the left and that the 600 Ma label is visible in the lower left hand corner), the continents were combined into one huge **supercontinent** that was mostly over what present-day ocean?
- ☐ Atlantic
- ☐ Pacific
- ☐ Indian
- ☐ Arctic

(b) Check and double-click placemark Problem 1b. Approximately 460-470 Ma, were the landmasses predominantly in the northern or southern hemisphere?
- ☐ northern
- ☐ southern

(c) Check and double-click placemark Problem 1c. In the early Paleozoic, vast areas of the continents were flooded with shallow seas called **epicontinental seas** (420-430 Ma, for example) where life flourished. How are these areas characterized on the paleogeographic maps?
- ☐ brown areas
- ☐ dark blue areas
- ☐ light blue areas

(d) Check and double-click placemark Problem 1d. Watch the landmass in the center of the view as the animation goes from Problem 1c to 1d (do this a couple of times to study it carefully). This landmass will become North America. During this time period, it is growing by accreting terranes to its margins. What kind of plate tectonic boundary is likely responsible for these series of mountain-building events?
- ☐ transform
- ☐ divergent
- ☐ convergent

(e) Check and double-click placemark Problem 1e. A major continent-continent collision is occurring to form what **supercontinent** at around 270-280 Ma?
- ☐ Rodinia
- ☐ Pangaea
- ☐ Laurentia
- ☐ Gondwana

(f) Check and double-click placemark Problem 1f. What mountain belt does this convergence form (*note that the eastern coast of North America figures prominently in this collision*)?

☐ Himalayas

☐ Ediacarans

☐ Mazatzals

☐ Appalachians/Caledonides

(g) Check and double-click placemark Problem 1g. When did the supercontinent start rifting apart?

☐ 350-300 Ma

☐ 280-250 Ma

☐ 220-170 Ma

☐ 100-50 Ma

(h) Check and double-click placemark Problem 1h. When did South America and Africa begin to separate?

☐ 170-105 Ma

☐ 220-180 Ma

☐ 100-50 Ma

☐ 50-25 Ma

(i) Check and double-click placemark Problem 1i. Approximately how long did it take India to travel from Antarctica to collide with Asia? *Start at 120 Ma and progressively drag the time slider until India collides with Asia.*

☐ ~150 m.y.

☐ ~30 m.y.

☐ ~200 m.y.

☐ ~90 m.y.

Name:_____

Energy & Mineral Resources

To answer questions for this worksheet, go to the following Geotours folder in Google Earth:
2. Exploring Geology Using Google Earth > Geotour Worksheets > L. Energy & Mineral Resources

L. Energy & Mineral Resources Worksheet

Supplemental background material also may be available in your textbook, through various internet resources, and within files in the 2. Exploring Geology Using Google Earth > Geotour Site Library.

Hydrocarbon Resources
- Worldwide map of major oil reserves
- Gulf Coast, USA

Coal Resources
- Farmersburg, IN

Other Energy Resources
- Rockford, IL

Mineral Resources
- Bingham Copper Mine, UT

Energy & Mineral Resources

Hydrocarbon Resources

1. Check the "Major Known Oil Reserves" folder in the **L. Energy & Mineral Resources** worksheet folder and make the entire folder semi-transparent. Also, turn on *Borders and Labels* in the Layers panel.

(a) Which of the following regions is <u>not</u> a location that has major oil reserves?
☐ Gulf of Mexico
☐ Saudi Arabia
☐ Japan
☐ Niger Delta, Africa *(in the bend on the west coast of Africa)*

(b) Which continent presently does <u>not</u> have any major known oil reserves?
☐ Asia
☐ Africa
☐ Antarctica
☐ South America

(c) Movement of subsurface salt deposits produces complex structures. In some places salt flows up to form **salt domes** and **salt anticlines**. When this happens, the salt that flows into domes must be withdrawn from somewhere else. In places where the salt is withdrawn, the overlying beds sink down to form a **basin**. Hydrocarbons sometimes get trapped in reservoirs adjacent to the margins of domes. Check and double-click the placemarks for Problems 1c-i and -ii. Which feature does each placemark point to?
☐ Problem 1c-i: basin and Problem 1c-ii: dome/anticline
☐ Problem 1c-ii: basin and Problem 1c-i: dome/anticline

(d) Check and double-click the placemark for Problem 1d. What kind of structure is this?
☐ dome
☐ basin

(e) To get a sense of scale of this feature, measure its diameter using the Ruler Tool 🔲 (in km between placemarks Problem 1e). What is its diameter?
☐ 5-15 km
☐ 20-30 km
☐ 0.1-0.5 km
☐ 0.01-0.05 km

Coal Resources

2. Check and double-click placemark Problem 2 to fly to a surface coal strip mine near Farmersburg, IN.

(a) Miners refill the active pit after coal has been extracted, then spread topsoil over the area and replant it. Nevertheless, reclaimed land will not look like unmined land for quite some time. Which placemark identifies a region that has been reclaimed?

☐ Problem 2a-i

☐ Problem 2a-ii

☐ Problem 2a-iii

(b) Extracting coal in open-pit strip mines may lead to unwanted consequences. When water reacts with minerals in excavated rock, the runoff into lakes and streams contains elements that affect water quality. Check and double-click placemark Problem 2b (also fly over the mine and nearby reclaimed areas). What evidence do you see that suggests possible contamination?

☐ water in the ponds has an unusual green color

☐ orange sediments suggest iron oxidation

☐ both of the first two statements

Other Energy Sources

3. In addition to hydrocarbons and coal, there are other types of energy sources as well.

(a) Check and double-click placemark Problem 3a to fly to an area south of Rockford, IL. What type of energy source is shown here?

☐ windmill farm

☐ hydroelectric

☐ solar farm

☐ geothermal

(b) Is this type of energy "green" (renewable, non-polluting) or "not green" (it uses non-renewable resources and is possibly polluting the environment)?

☐ green

☐ not green

(c) Check and double-click placemark Problem 3c to fly to an area NW of Rockford, IL. What type of energy source is shown here?

☐ windmill farm

☐ hydroelectric

☐ nuclear

☐ solar farm

(d) Check and double-click placemark Problem 3d to zoom out from this area to an altitude of about 250 km. Assuming that weather systems follow a general path from west to east, what major metropolitan area is downwind of this facility?

☐ Madison, WI

☐ Milwaukee, WI

☐ Quad Cities (Rock Island, IL; Moline, IL; Davenport, IA; and Bettendorf, IA)

☐ Chicago, IL

Mineral Resources

4. Mineral resources that we use in many everyday items must be extracted from the earth as well.

(a) Check and double-click the placemarks for Problem 4a to fly to Bingham Copper Mine, the largest open-pit mine in the world. Using the Ruler Tool [▯] (in km between placemarks Problem 4a), estimate the diameter of the main pit.
- ☐ 11-15 km
- ☐ 3-5 km
- ☐ 0-1 km
- ☐ 21-25 km

(b) Check and double-click the placemarks for Problem 4b and use the Hand cursor to measure the elevation difference (in m) between the two placemarks to get an estimate for the depth of the main pit.
- ☐ 600-700 m
- ☐ 1200-1300 m
- ☐ 250-350 m
- ☐ 80-140 m

(c) Check and double-click placemark Problem 4c to visit one of the many tailings/waste piles. What evidence exists in this view to suggest that this process has similar environmental issues to the coal strip mine in Indiana?
- ☐ water ponds have discolored water
- ☐ reclaimed areas have only small trees on them

Mass Movements

To answer questions for this worksheet, go to the following Geotours folder in Google Earth:
2. Exploring Geology Using Google Earth > Geotour Worksheets > M. Mass Movements

M. Mass

Movements

Worksheet

Supplemental background material also may be available in your textbook, through various internet resources, and within files in the 2. Exploring Geology Using Google Earth > Geotour Site Library.

Portuguese Bend Landslide, CA
- Portuguese Bend, CA

La Conchita Mudslide, CA
- La Conchita, CA

Gros Ventre Slide, WY
- Gros Ventre Wilderness, WY

Vaiont Dam Slide, Italy
- Vaiont River/Reservoir, Italy

Geotour Worksheet M

Mass Movements

Portuguese Bend Landslide, CA

1. Double-click placemark Problem 1 in the **M. Mass Movements** worksheet folder. This flies you to the infamous Portuguese Bend area in California. This region has experienced a massive (and progressive) landslide that has cannibalized the subdivision that once existed on the cliffs overlooking the ocean.

(a) Check and double-click placemark Problem 1a to fly in closer to the region. Many of these buildings reside on a hummocky debris apron. How did urbanization likely contribute to cause the mass wasting? *You may want to use your textbook.*

☐ water from septic systems, sprinklers, etc. lubricated the slope making it more unstable

☐ undercutting the slope to make flatter areas for buildings made the slope more unstable

☐ weight of buildings and roads made the area more unstable

☐ all of the above

(b) Check and double-click placemark Problem 1b. Do you think making the road here was a smart thing to do?

☐ yes, it will help stabilize the slope

☐ yes, it will serve as a boundary to prevent further building on the unstable area

☐ no, the road just undercuts the slope above it, making it even more unstable

(c) Would you buy the home highlighted by placemark Problem 1c?

☐ no, it is going to experience additional collapse

☐ yes, it will likely be cheap, and no problems will happen in my lifetime

☐ yes, it will likely be cheap, and I can make a quick profit

La Conchita Mudslide, CA

2. When you double-click on placemark Problem 2a, you will be looking ENE at the steep, seaward-sloping face behind the flat, coastal sea terrace on which people have built homes. The placemark is positioned on the most recent, active mudslide, which is the infamous La Conchita mudslide.

(a) Why do you think this mudslide occurred?

☐ the area has little vegetation to hold the soil

☐ the area could have been undercut by the road half-way up the slope

☐ the sea terrace on which the houses sit has over-steepened the cliff

☐ all of the above

(b) What do the areas marked by placemarks Problem 2b represent?

☐ construction sites

☐ older mudslides

☐ former tsunami erosional features

(c) Do you want to buy a house near placemark Problem 2c (check and double-click it)?

☐ yes, it will likely be cheap, and no problems will happen in my lifetime

☐ no, it is likely to be destroyed by a mudslide like the La Conchita mudslide

☐ yes, it will likely be cheap, and I can make a quick profit

Gros Ventre Slide, WY

3. Check and double-click on placemark Problem 3a, which will fly you to a view of the Gros Ventre Slide and the adjacent Slide Lake.

(a) What specific part of the slide is placemark Problem 3a?

☐ a scar at the "head" of the slide where trees and soil have been stripped away

☐ a hummocky debris apron of material at the "toe" of the slide

☐ the location has not experienced any mass movement

(b) Check and double-click on placemark Problem 3b. Based on the position of the lake, in what direction does the water of the Gros Ventre River flow at this locality?

☐ west to east

☐ east to west

(c) Which is not a reason that might have contributed to the slide?

☐ river erosion had made the slope unstable

☐ heavy rains made the materials heavier and lubricated the slip surface

☐ rock layers were dipping towards the valley

☐ a large housing development was built in the area

Vaiont Dam Slide, Italy

4. Check and double-click on placemark Problem 4 to fly you to the site of the Vaiont Dam disaster that occurred in 1963.

(a) Which answer correctly identifies the features highlighted by placemarks Problem 4a-i, -ii, and -iii?

☐ 4a-i: dam; 4a-ii: hummocky debris apron; and 4a-iii: slide scar

☐ 4a-i: hummocky debris apron; 4a-ii: slide scar; and 4a-iii: dam

☐ 4a-i: slide scar; 4a-ii: dam; and 4a-iii: hummocky debris apron

(b) Check and double-click on placemark Problem 4b. Which is not a reason that might have contributed to the slide?

☐ the layers dip towards the valley

☐ the reservoir formed by the dam lubricated the slip surface

☐ the weight of buildings in the Italian village of Casso (located at placemark Problem 4b)

Name:_____

Stream Landscapes

To answer questions for this worksheet, go to the following Geotours folder in Google Earth:
2. Exploring Geology Using Google Earth > Geotour Worksheets > N. Stream Landscapes

Supplemental background material also may be available in your textbook, through various internet resources, and within files in the 2. Exploring Geology Using Google Earth > Geotour Site Library.

N. Stream

Landscapes

Worksheet

Headward Erosion
- Canyonlands National Park, UT

Stream Patterns
- Mesa Verde National Park, CO
- Appalachian Valley & Ridge, PA
- Montrose, CO
- Northeastern AZ

Meandering Stream Features
- Rio Ucayali, Peru
- Mississippi River near Vicksburg, MS
- Campti, LA
- Goosenecks State Park, UT
- Green River near Bowknot Bend, UT

Miscellaneous Stream Features
- Kaaterskill, NY
- Sheep Mountain, WY

Geotour Worksheet **N**

Stream Landscapes

Headward Erosion

1. Check and double-click placemark Problem 1a in the **N. Stream Landscapes** worksheet folder. In this view from Grand View Point in Canyonlands National Park, three stream branches (the middle one is labeled with the placemark) are experiencing headward erosion. Headward erosion allows streams to lengthen in the upstream direction.

(a) As seen from this viewpoint, in which direction (right or left) are the streams lengthening?

☐ right (*west*)

☐ left (*east*)

(b) Check and double-click placemark Problem 1b. In which direction are the streams experiencing <u>headward erosion</u>?

☐ NE

☐ SW

☐ NW

☐ SE

Stream Patterns

2. Streams can develop a variety of patterns depending on the rock type and structure that underlies them. *See your textbook for diagrams of stream patterns.*

(a) What is the stream pattern exhibited at placemark Problem 2a?

☐ trellis *(streams meet at right angles)*

☐ rectangular *(streams have right-angle bends)*

☐ radial *(streams radiate from a central high point)*

☐ dendritic *(streams resemble veins in a leaf)*

(b) Check and double-click the folder Problem 2b. What is the <u>overall general</u> stream pattern exhibited at this location with tributaries flowing in to the main trunk stream *(see blue lines)*?

☐ trellis *(streams meet at right angles)*

☐ parallel *(streams are approximately parallel to each other)*

☐ radial *(streams radiate from a central high point)*

☐ dendritic *(streams resemble veins in a leaf)*

(c) Leaving folder Problem 2b checked, check the placemarks for Problem 2c. Although the regional network of streams in the Appalachian Valley and Ridge Province is the pattern identified in Problem 2b, other stream patterns may develop locally. Note the angle at which the tributaries labeled by the placemarks for Problem 2c intersect other streams, and note the angle of bends in these streams. Based on these observations, what is the local drainage pattern for the streams defined by the placemarks for Problem 2c?

☐ parallel *(streams are approximately parallel to each other)*
☐ rectangular *(streams have right-angle bends)*
☐ radial *(streams radiate from a central high point)*
☐ dendritic *(streams resemble veins in a leaf)*

(d) What is the stream pattern exhibited at placemark Problem 2d?

☐ trellis *(streams meet at right angles)*
☐ parallel *(streams are approximately parallel to each other)*
☐ rectangular *(streams have right-angle bends)*
☐ radial *(streams radiate from a central high point)*

(e) What is the stream pattern exhibited at placemark Problem 2e?

☐ trellis *(streams meet at right angles)*
☐ parallel *(streams are approximately parallel to each other)*
☐ rectangular *(streams have right-angle bends)*
☐ radial *(streams radiate from a central high point)*

Meandering Stream Features

3. Meandering streams develop various landforms along the stream bed and its floodplain.

(a) What is the trace of a former stream meander called (placemark Problem 3a)?

☐ cutbank
☐ point bar
☐ yazoo tributary
☐ meander scar

(b) What is the outside bend of a meander called (placemark Problem 3b)?

☐ cutbank
☐ point bar
☐ oxbow lake
☐ yazoo tributary

(c) What is the deposit on the inside bend of a meander called (placemark Problem 3c)?

☐ cutbank
☐ point bar
☐ natural levee
☐ meander scar

(d) What is the stream mostly flowing in the larger river's floodplain called (placemark Problem 3d)?

☐ point bar
☐ oxbow lake
☐ yazoo tributary
☐ meander scar

(e) What is an abandoned meander with water called (placemark Problem 3e)?
- ☐ cutbank
- ☐ oxbow lake
- ☐ yazoo tributary
- ☐ natural levee

(f) As you have seen, most meandering streams occupy broad, flat floodplains. What has happened to the meandering stream shown by placemark Problem 3f?
- ☐ it has become incised/entrenched because of a change in gradient or base level
- ☐ it has carved a steep canyon because it experienced a catastrophic flood
- ☐ it is a normal meandering stream that has just experienced a lot of mass movements

(g) Check and double-click placemark Problem 3g. This meander has been cutoff sometime in the past and now has been by-passed by the Green River. Did the cutoff happen before or after the river incised/entrenched into the landscape?
- ☐ after, because the bend has been incised/entrenched as well
- ☐ before, the river could not have cut through the meander neck if it was already entrenched

---

Miscellaneous Stream Features

4. There are many other fascinating features about streams...here, we'll highlight two more.

(a) Check and double-click the "Stream Piracy, Kaaterskill, NY" folder. Why are the dark blue and orange streams **pirating/capturing** the headwaters of the cyan and magenta streams?
- ☐ they are larger and can hold more water
- ☐ they have steeper gradients and are eroding headwardly
- ☐ the land to the west is being uplifted
- ☐ landslides have diverted the water

(b) Check and double-click placemark Problem 4b. Notice that the stream cuts across the Sheep Mountain anticline instead of being diverted around it. Such streams are generally called **antecedent streams** or **superposed streams**. **Antecedent streams** are generally older than the structure and continue their downcutting at a pace similar to the rate at which the fold is *actively growing*. In contrast, **superposed streams** develop long after a structure has developed and been buried by sediment. The younger stream develops its stream valley and cuts through the older structure as it exhumes it. In the Layers panel, turn on *Gallery > Earthquakes*. Given present day earthquake activity in this area, what is the most likely interpretation for the stream?
- ☐ it is antecedent
- ☐ it is superposed

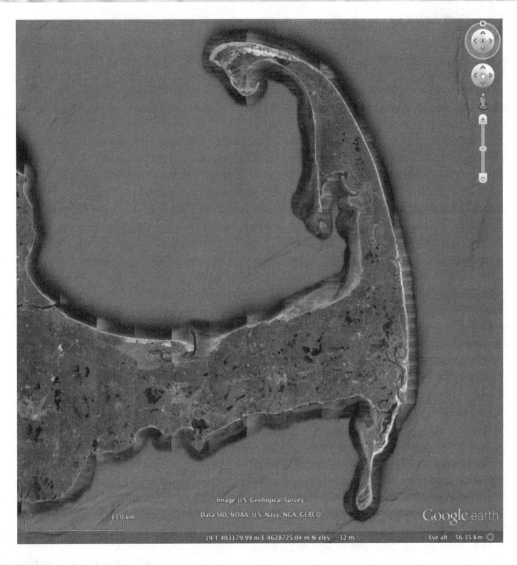

To answer questions for this worksheet, go to the following Geotours folder in Google Earth:
2. Exploring Geology Using Google Earth > Geotour Worksheets > O. Oceans & Coastlines

Supplemental background material also may be available in your textbook, through various internet resources, and within files in the 2. Exploring Geology Using Google Earth > Geotour Site Library.

O. Oceans &

Coastlines

Worksheet

Seafloor Bathymetry
- Southern margin of South America

Coral Reefs
- Various islands, South Pacific

Barrier Islands & Spits
- Cape Hatteras, NC
- Cape Cod, MA
- Lake Michigan, Chicago, IL

Coastlines & Sea-Level Rise
- East coast, USA

Oceans & Coastlines

Sea-Floor Bathymetry

1. As you have seen in previous chapters, the sea-floor is not flat and featureless, but rather has numerous interesting landforms that reveal much about geologic processes active on our planet.

(a) Check and double-click the placemarks for Problems 1a-i, -ii, and -iii in the **O. Oceans & Coastlines** worksheet folder to fly to the southern tip of South America. Identify the bathymetric feature associated with each placemark. *See your textbook if you forget what these things are.*

☐ 1a-i: continental shelf; 1a-ii: continental slope; and 1a-iii: abyssal plain

☐ 1a-i: continental slope; 1a-ii: continental shelf; and 1a-iii: abyssal plain

☐ 1a-i: abyssal plain; 1a-ii: continental slope; and 1a-iii: continental shelf

☐ 1a-i: abyssal plain; 1a-ii: continental shelf; and 1a-iii: continental slope

(b) Check and double-click the placemarks for Problems 1b-i, -ii, and -iii. Identify the bathymetric feature associated with each placemark.

☐ 1b-i: trench; 1b-ii: mid-ocean ridge; and 1b-iii: transform fault

☐ 1b-i: mid-ocean ridge; 1b-ii: trench; and 1b-iii: transform fault

☐ 1b-i: mid-ocean ridge; 1b-ii: transform fault; and 1b-iii: trench

☐ 1b-i: transform fault; 1b-ii: trench; and 1b-iii: mid-ocean ridge

(c) Check and double-click the placemarks for Problem 1c-i and 1c-ii. Which placemark shows bathymetry associated with an **active margin** (associated with an active plate boundary) and a **passive margin** (no nearby active plate boundary)?

☐ 1c-i: active margin and 1-ii: passive margin

☐ 1c-i: passive margin and 1-ii: active margin

Coral Reefs

2. Coral reefs are communities of organisms that can build spectacular landforms in relatively shallow water depths (<60 m) at low latitudes (<30°).

(a) Check and double-click placemark Problem 2a to fly to some scenic islands in the South Pacific. What kind of coral reef is surrounding the island?

☐ barrier reef *(forms offshore from the coast with an intervening lagoon)*

☐ fringing reef *(forms directly along the coast)*

☐ atoll *(forms around a circular reef surrounding a lagoon)*

(b) Check and double-click placemark Problem 2b to fly to a nearby island in the South Pacific. What kind of coral reef is surrounding the island?

☐ barrier reef *(forms offshore from the coast with an intervening lagoon)*

☐ fringing reef *(forms directly along the coast)*

☐ atoll *(forms around a circular reef surrounding a lagoon)*

(c) Check and double-click placemark Problem 2c to fly to another nearby area. What kind of coral reef is present here?

☐ barrier reef *(forms offshore from the coast with an intervening lagoon)*

☐ fringing reef *(forms directly along the coast)*

☐ atoll *(forms around a circular reef surrounding a lagoon)*

Barrier Islands & Spits

3. Barrier islands, beaches, and spits form in coastal areas with abundant sand.
(a) Check and double-click placemark Problem 3a to fly to Cape Hatteras off the coast of North Carolina. Use the Hand Tool to determine the range of heights for the sand ridge specified by the placemark (presumably this sand ridge would provide protection from waves of this height or lower for areas on the lagoon side of the ridge).
    - [ ] 0-1m
    - [ ] 3-10m
    - [ ] 15-20m
(b) Why is most of the construction on the lagoon side of the barrier island (placemark Problem 3b)?
    - [ ] more sunlight
    - [ ] protected from storm waves and erosion
    - [ ] land there is at a higher elevation
    - [ ] there are more beach sands on that side
(c) Check and double-click the placemarks for Problem 3c to fly to Cape Cod, MA. Here, the remains of an E-W oriented glacial moraine are being reshaped into a spit by a longshore current. What is the general direction of the current (*look at which way was material moved*)?
    - [ ] from E to W
    - [ ] from W to E
    - [ ] from N to S
    - [ ] from S to N
(d) Piers or groins have been built out into Lake Michigan in Chicago to retain sand for beaches and to prevent sand erosion (placemark Problem 3d). Which direction is the current that is moving the sand? *The current direction is from the narrow part of the sand "triangle" towards the wider part.*
    - [ ] from E to W
    - [ ] from W to E
    - [ ] from N to S
    - [ ] from S to N

Coastlines & Sea-Level Rise

4. Turn on the "U.S. East Coast Sea Level Changes" overlay (make it semi-transparent) and double-click the placemarks for Problems 4a-i, -ii, -iii, and -iv in the Problem 4 folder.
(a) If the Greenland and Antarctica ice sheets melt, where would be the best place to own land?
    - [ ] Problem 4a-i
    - [ ] Problem 4a-ii
    - [ ] Problem 4a-iii
    - [ ] Problem 4a-iv

Groundwater &
Karst Landscapes

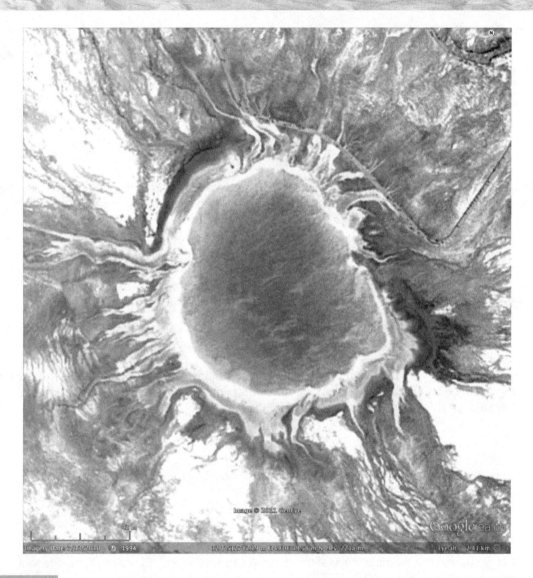

To answer questions for this worksheet, go to the following Geotours folder in Google Earth:
2. Exploring Geology Using Google Earth > Geotour Worksheets > P. Groundwater & Karst Landscapes

Supplemental background material also may be available in your textbook, through various internet resources, and within files in the 2. Exploring Geology Using Google Earth > Geotour Site Library.

P. Groundwater

& Karst

Landscapes

Worksheet

Irrigation in the Saudi Desert
- Desert irrigation, Saudi Arabia

Surface and Groundwater Flow
- Everglades National Park, FL

Hot Springs
- Yellowstone National Park, WY

Karst Features
- Mammoth Cave National Park, KY
- Orleans, IN
- Nalingxiang, China
- Winter Park, FL

77

Geotour Worksheet P

Groundwater & Karst Landscapes

Irrigation in the Saudi Desert

1. Check and double-click polygon Problem 1 in the Problem 1 folder within the **P. Groundwater & Karst Landscapes** worksheet folder to fly to a region in Saudi Arabia.

(a) What is the trend/orientation of the longest dimension of the polygon (and also of the water wells and irrigation circles)?
- ☐ NW-SE
- ☐ NE-SW
- ☐ N-S
- ☐ E-W

(b) Given that nearby ridges also have this same trend, what is the best explanation for why the water wells and irrigation circles have this trend and are only located in certain "bands"?
- ☐ they just haven't gotten around to drilling the other areas
- ☐ the water may be contained in **aquifers** that have this trend
- ☐ the other areas receive plenty of surface water and don't require well water

Surface and Groundwater Flow

2. Surface water flow is influenced most directly by elevation differences. Groundwater flow is controlled by elevation differences in the water table, which commonly reflects surface elevations to a degree.

(a) Use the Hand Tool to determine the elevations at the placemarks for Problems 2a-i and 2a-ii. What direction should the surface and groundwater flow? (*Confirm this by turning on the "Everglades, FL-Previous Groundwater Flow" overlay*).
- ☐ N to S
- ☐ S to N
- ☐ E to W

(b) Turn on the "Everglades, FL-Present Groundwater Flow" overlay. What has changed?
- ☐ part of the swamps have been drained
- ☐ canals have diverted water resources to coastal cities
- ☐ groundwater flow directions have been changed by the canal system
- ☐ all three of the above answers are correct

(c) What is causing the dramatic saltwater incursion as seen on the "Everglades, FL-Present Groundwater Flow" overlay?
- ☐ salt water is less dense than fresh water
- ☐ various urbanization changes (e.g., pumping freshwater from wells) affecting the slope of the water table under the land
- ☐ sea level fall

Hot Springs

3. Check and double-click the placemarks for Problem 3a. You will fly to Midway Geyser Basin and see Grand Prismatic Pool, a hot spring, and the nearby Firehole River in Yellowstone NP.

(a) Use the Ruler Tool [image] to measure the width of Grand Prismatic Pool (in m) between the placemarks of Problem 3a.
- ☐ 125-135 m
- ☐ 291-301 m
- ☐ 266-276 m
- ☐ 980-990 m

(b) Turn on placemarks Problem 3b. Using the Hand Tool, compare the elevation of Grand Prismatic Pool and of the Firehole River (in m). Does overland flow of water travel from the pool to the Firehole River or from the Firehole River to Grand Prismatic Pool?
- ☐ Firehole River to Grand Prismatic Pool
- ☐ Grand Prismatic Pool to Firehole River

(c) Check and double-click placemark Problem 3c to fly to Mammoth Hot Springs, an area in the NW corner of Yellowstone National Park. This area is different than the silica-based geyser features to the south that are associated with volcanic rocks. Here, hot groundwater circulates in the subsurface to dissolve minerals from the Madison Limestone. What are these terraces of reprecipitated minerals made of (recall what mineral typically forms limestone)?
- ☐ quartz
- ☐ calcite
- ☐ gypsum
- ☐ pyrite

Karst Features

4. Acidic groundwater can dissolve soluble rocks to form fascinating **karst landscapes**.

(a) Check and double-click placemark Problem 4a to fly to the Mammoth Cave, KY area. This flat area is known as the **Pennyroyal Plateau**. What are the depressions (often filled with water or clumps of trees) that dot the landscape called?
- ☐ dry holes
- ☐ springs
- ☐ sinkholes
- ☐ caves

(b) Check and double-click placemarks Problem 4b to fly to an area near Orleans, IN. The eastern placemark points to the Lost River, a stream that flows from east to west in this area. Follow the stream to the west to the western placemark. What kind of stream is it?
- ☐ intermittent
- ☐ disappearing/sinking
- ☐ arroyo
- ☐ perennial

(c) The Lost River really isn't "lost". West of placemarks Problem 4b, we see glimpses of the water that has been diverted underground in areas like Elrod Gulf (placemark Problem 4c). Such areas are called **karst windows**, as they provide a "window" into the groundwater system. Eventually, the groundwater resurfaces at a **resurgent spring** near Orangeville and the Lost River flows at the surface again. How is the area between placemarks Problem 4b best described?

☐ an area of impermeable rock (e.g., shale)

☐ an area of tower karst

☐ a sinkhole plain of permeable limestone that only allows surface flow in very wet periods

(d) What kind of karst landscape is shown in the area around placemark Problem 4d?

☐ tower karst

☐ sinkhole plain

☐ karst window

(e) What kind of karst feature is shown by placemark Problem 4e (see photo in the placemark)?

☐ cave

☐ monadnock

☐ dry valley

☐ sinkhole

Geotour Worksheet Q

Desert Landscapes

To answer questions for this worksheet, go to the following Geotours folder in Google Earth:
2. Exploring Geology Using Google Earth > Geotour Worksheets > Q. Desert Landscapes

Supplemental background material also may be available in your textbook, through various internet resources, and within files in the 2. Exploring Geology Using Google Earth > Geotour Site Library.

Q. Desert

Landscapes

Worksheet

Desert and Desert Features
- Giza, Egypt
- Chala, Peru
- Tarim Basin, China
- Uluru (Ayers Rock), Australia
- Death Valley, CA
- Monument Valley, AZ

Water Use in Arid Regions
- Phoenix, AZ
- Aral Sea, Central Asia

Geotour Worksheet Q

Desert Landscapes

Deserts and Desert Features

1. A **desert** is a region that is so arid that it contains no permanent streams, except for those that bring water in from temperate regions elsewhere.

(a) Check and double-click placemark Problem 1a in the **Q. Desert Landscapes** worksheet folder to fly to the Pyramids of Giza. These structures are located in the largest desert in the world (the desert covers most of northern Africa). What is the name of this desert?
- ☐ Gobi
- ☐ Sahara
- ☐ Mojave
- ☐ Kalahari

(b) Which desert is the driest place on Earth? Check and double-click placemark Problem 1b to fly to this desert along the western coast of South America.
- ☐ Gobi
- ☐ Sahara
- ☐ Atacama
- ☐ Mojave

(c) Check and double-click placemark Problem 1c. What direction did the wind blow to create this **Barchan dune**? *Hint: Look at the tips of the dune; they form downwind.*
- ☐ S to N
- ☐ N to S

(d) Check and double-click the Tarim Basin polygon to fly to this desert in China. Use the Ruler Tool  to estimate the longest axis of the Tarim Basin polygon (km).
- ☐ 200-300 km
- ☐ 900-1000 km
- ☐ 50-150 km
- ☐ 1200-1400 km

(e) Check and double-click placemark Problem 1e. Why does this north-flowing intermittent stream terminate in the middle of the desert, even though the land slopes gently north?
- ☐ it disappears into a sinkhole
- ☐ evaporation and infiltration are too great
- ☐ it has been dammed up to create a reservoir

(f) Check and double-click the placemarks for Problem 1f to fly to Ayers Rock in Australia. Use the Hand Tool to determine how high this rock stands above its surroundings (in m, measure the placemarks).
- ☐ 500-530 m
- ☐ 90-130 m
- ☐ 300-350 m
- ☐ 800-850 m

(g) Ayers Rock is an erosional remnant of one limb of a regional syncline (the lines across the rock are steeply dipping beds). What term is used to described such resistant erosional remnants that stand out in relief above the alluvium in arid settings?
- ☐ klippe
- ☐ roche moutonee
- ☐ inselberg
- ☐ horn

(h) Check and double-click placemark Problem 1h to fly to Death Valley, CA. What is the highlighted feature that forms evaporite deposits in arid settings?
- ☐ playa lake
- ☐ bajada
- ☐ butte
- ☐ inselberg

(i) Check and double-click placemarks Problem 1i. What are the highlighted features?
- ☐ alluvial fans
- ☐ mesas
- ☐ buttes
- ☐ inselbergs

(j) Check and double-click placemark Problem 1j. What is the highlighted feature?
- ☐ mesa
- ☐ butte

(k) Check and double-click placemark Problem 1k. What is the highlighted feature?
- ☐ mesa
- ☐ butte

Water Use in Arid Regions

2. Water usage in arid regions is becoming an increasingly important issue.

(a) Check and double-click the placemarks for Problems 2a-i and 2a-ii to fly to the Phoenix, AZ area. How are the areas highlighted by the placemarks different?
- ☐ 2a-i: lush, green golf course and 2a-ii: arid desert with dry washes
- ☐ 2a-i: arid desert with dry washes and 2a-ii: lush, green golf course

(b) What are the features highlighted by the placemarks for Problem 2b (click all seven!)?
- ☐ landslides
- ☐ marinas
- ☐ dams
- ☐ wildlife sanctuaries

(c) What is the feature highlighted by placemark Problem 2c?
- ☐ interstate
- ☐ water canal
- ☐ railroad
- ☐ bike path

(d) Follow the feature in Problem 2c to the west (continue across the tunnels/pipelines that look like gaps). Where does it ultimately originate (the placemarks for Problem 2d mark the ends of a long tunnel)?

☐ Colorado River

☐ Pacific Ocean

☐ a glacier in the Sierra Nevada Mountains

☐ Gulf of California

Just for Fun..._Considering your answers to the questions above and to their arid location, are Phoenix and the surrounding communities sustainable? Can they grow larger? Should they grow larger?_

(e) Check and double-click the "Aral Sea, Central Asia" historical animation. Click the historical imagery icon in the toolbar and pan through the historical imagery of the Aral Sea as water has been diverted for human uses instead of flowing into the sea. Comparing the 1973 imagery to the most recent 2010 imagery, what happened to the sea?

☐ the sea became significantly larger

☐ the sea dramatically shrank in size

☐ there was very little change in the sea

Name:_____

Geotour Worksheet R

Glacial Landscapes

To answer questions for this worksheet, go to the following Geotours folder in Google Earth:
2. Exploring Geology Using Google Earth > Geotour Worksheets > R. Glacial Landscapes

Supplemental background material also may be available in your textbook, through various internet resources, and within files in the 2. Exploring Geology Using Google Earth > Geotour Site Library.

R. Glacial

Landscapes

Worksheet

Continental Glacier Features
- Palmyra, NY
- Finger Lakes, NY
- Long Island, NY
- Cape Cod, MA

Alpine/Valley Glacial Features
- Baffin Island, Canada
- Matterhorn, Switzerland
- Sierra Nevada Mountains, CA
- Holy Cross, CO
- Mono Lake, CA

Piedmont Glacial Features
- Malaspina Glacier, AK

85

Glacial Landscapes

Continental Glacial Features

1. Continental glaciers created a wide array of fascinating landforms (many of which people pass by every day without noticing).

(a) Check and double-click placemark Problem 1a in the **R. Glacial Landscapes** worksheet folder to fly to northern New York state. Once there, temporarily increase your vertical exaggeration to 3. Now click the *Google Maps* button  in the icon bar. Once *Google Maps* opens in your web browser, click the *Map* icon (upper right-hand corner) and select *Terrain* from the menu, which allows you to see a shaded relief map with contour lines (lines of equal elevation). What are the numerous landforms that you see here?
- ☐ kettles
- ☐ drumlins/flutes

(b) Using the geometry of these features, <u>from what direction</u> did the continental glacier advance (originate)? *Glaciers generally advanced from the direction where the contour lines on the landforms are closest together (steep slope) towards the direction where the contour lines on the landforms are farther apart (gentle slope).*
- ☐ N
- ☐ S

(c) What are these features mostly composed of? *You may need to refer to your book.*
- ☐ ice-carved rock
- ☐ glacial till
- ☐ tightly folded anticlines & synclines

(d) Go back to *Google Earth* and check and double-click placemark Problem 1d to see the nearby Finger Lakes region of northern New York state. How did these lakes form?
- ☐ these were subglacial tunnels that channeled meltwater
- ☐ these formed from the melting of icebergs
- ☐ glaciers scoured deep grooves in the landscape that filled with water

(e) Check and double-click the "Glacial Moraines-Long Island, NY & Cape Cod, Ma" map overlay. What kind of moraines likely were deposited on Long Island & Cape Cod?
- ☐ recessional/terminal/end
- ☐ lateral
- ☐ medial
- ☐ ground

(f) Given your answer to Problem 1e, <u>from what direction</u> did the continental glacier advance?
- ☐ N
- ☐ S
- ☐ E
- ☐ W

Alpine/Valley Glacial Features

2. Alpine and valley glaciers sculpt the landscape over which they flow and create numerous spectacular landforms. The following problems will explore some of these in more detail.

(a) Check and double-click the placemarks for Problem 2a to fly to Baffin Island in Canada. What type of moraine do these placemarks highlight?
- ☐ ground
- ☐ lateral
- ☐ medial
- ☐ terminal/end

(b) Check and double-click placemark Problem 2b. What type of moraine does this placemark highlight?
- ☐ ground
- ☐ lateral
- ☐ medial
- ☐ terminal/end

(c) The glacier highlighted by placemark Problem 2c extends down the valley to actually have its terminus extend into the seawater of the fjord. What type of glacier is this?
- ☐ tidewater
- ☐ piedmont
- ☐ cirque

(d) Check and double-click the placemarks for Problem 2d to fly to the Matterhorn area in Switzerland. What narrow, knife-like feature that once separated two alpine glaciers do these placemarks highlight?
- ☐ cirque
- ☐ horn
- ☐ truncated spur
- ☐ arête

(e) Check and double-click the placemarks for Problem 2e. What semi-circular, amphitheater-shaped feature do these placemarks highlight?
- ☐ cirque
- ☐ horn
- ☐ truncated spur
- ☐ arête

(f) Check and double-click the placemarks for Problem 2f. What jagged feature that is a result of glaciers sculpting the rock on 3 or more sides do these placemarks highlight?
- ☐ cirque
- ☐ horn
- ☐ truncated spur
- ☐ arête

(g) Check and double-click placemark Problem 2g. What saddle-like feature along the narrow, knife-like ridge does this placemark highlight?
- ☐ tarn
- ☐ col on an arête
- ☐ truncated spur
- ☐ hanging valley

(h) Check and double-click placemark Problem 2h to fly to the Sierra Nevada Mountains in California. What feature does this placemark highlight?
- ☐ cirque
- ☐ horn
- ☐ col on an arête
- ☐ arête

(i) Check and double-click placemark Problem 2i. What feature does this placemark highlight?
- ☐ cirque
- ☐ horn
- ☐ col on an arête
- ☐ truncated spur

(j) Check and double-click placemark Problem 2j. What feature does this placemark highlight?
- ☐ cirque
- ☐ horn
- ☐ col on an arête
- ☐ arête

(k) Check and double-click placemark Problem 2k. What feature does this placemark highlight?
- ☐ cirque
- ☐ horn
- ☐ col on an arête
- ☐ arête

(l) Check and double-click placemark Problem 2l to fly to Holy Cross, CO. What feature dams the end of this glacially carved valley?
- ☐ terminal/end moraine
- ☐ lateral moraine
- ☐ medial moraine
- ☐ an arête

(m) Check and double-click the folder for Problem 2m to fly to Mono Lake, CA. All of the placemarks highlight the same type of feature. What are these features?
- ☐ terminal/end moraines
- ☐ lateral moraines
- ☐ ground moraines
- ☐ medial moraines

(n) Check and double-click the placemarks for Problems 2n-i and 2n-ii. Using cross-cutting relationships, which moraine is the youngest? *Recall that younger features truncate and cut across older features.*
- ☐ 2n-i
- ☐ 2n-ii

Piedmont Glacial Features

3. Sometimes alpine and valley glaciers spill out from an ice field/cap and coalesce when the topography becomes less steep, forming **piedmont glaciers** and associated landforms.

(a) Check and double-click placemark Problem 3a to fly to the Malaspina Glacier area in Alaska. What are the folds comprised of?

☐ layers of moraine and ice

☐ layers of solid rock

(b) Turn on the path labeled "Problem 3" to show the approximate boundary between rock and ice. What is causing the layers to be bent & distorted?

☐ plate tectonic convergence

☐ movement along a fault

☐ frictional drag as the glacial ice flows past the rocks to the right/east of the Problem 3 path

☐ this distortion is only apparent, the materials were actually deposited this way

(c) Check and double-click placemark Problem 3c. What circular water-filled depression does this placemark highlight?

☐ kettle

☐ kame

(d) Check and double-click placemark Problem 3d. What feature does this placemark highlight?

☐ braided stream on an outwash plain

☐ esker

Name:_____

Geotour Worksheet S

Global Change

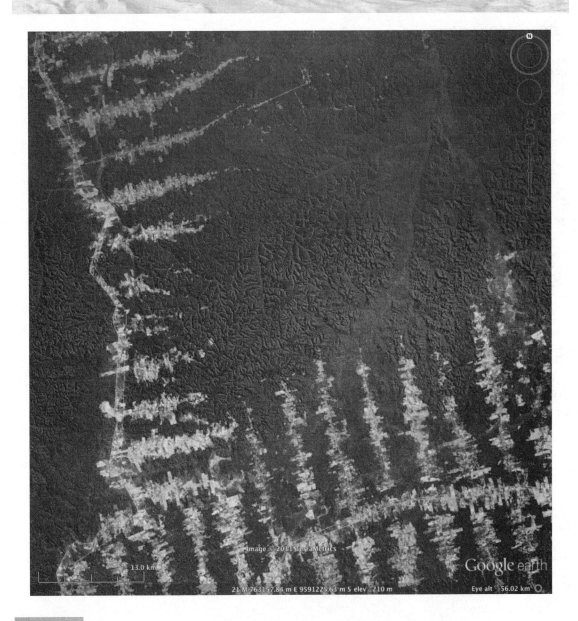

S. Global Change Worksheet

To answer questions for this worksheet, go to the following Geotours folder in Google Earth:
2. Exploring Geology Using Google Earth > Geotour Worksheets > S. Global Change

Supplemental background material also may be available in your textbook, through various internet resources, and within files in the 2. Exploring Geology Using Google Earth > Geotour Site Library.

Effects of Global Warming
- Larsen B ice shelf, Antarctica

Effects of Deforestation
- Deforestation, Rondonia, South America

Water Use in Arid Regions
- Phoenix, AZ

Preservation of Natural Habitats
- Everglades National Park, FL

91

Name:_____

Geotour Worksheet S

Global Change

Effects of Global Warming

1. One of the present global concerns is the breakup and melting of glaciers and ice shelves due to global warming.

(a) Check and double-click placemark Problem 1a in the **S. Global Change** worksheet folder to fly to Antarctica to view the area around the Larsen B ice shelf. Turn on the "Larsen B Ice Shelf" polygon. Once there, click the clock in the icon bar to look at historical imagery, focusing on imagery of Jan 31, 2002, Mar 07, 2002, & Feb 23, 2006 (*note: the Dec 1999 imagery is presently incorrect at the time of this publication*). What happened?

☐ the ice shelf disintegrated

☐ the ice shelf dramatically expanded

(b) To get a sense of scale, use the Ruler Tool ▮ to estimate the size of the magenta polygon (in km) using placemarks Problem 1b.

☐ 10-20 km

☐ 40-60 km

☐ 300-325 km

☐ 100-120 km

Effects of Deforestation

2. Humans also are modifying their environment in various ways (including deforestation) and dramatically affecting both local and global ecosystems.

(a) Check and double-click placemark Problem 2a to fly to South America to view deforestation over time and its effects on the landscape. Once there, click the clock in the icon bar to look at historical imagery, beginning with Jun 1975 and ending with Aug 2010. Estimate the percentage of land in the field of view that has experienced deforestation.

☐ 1%

☐ 15%

☐ 75%

(b) Turn on the "Deforestation Area" polygon and make it semi-transparent. This polygon represents an area of approximately 5000 sq km. Using your answer from Problem 2a, what area (in sq km) within the polygon has been deforested from 1975-2010?

☐ 500 sq km

☐ 1250 sq km

☐ 3750 sq km

☐ 4750 sq km

(c) On average, based on your answer to 2b, how much deforestation occurred in this area per year (sq km/yr)?

☐ 15 sq km/yr

☐ 38 sq km/yr

☐ 114 sq km/yr

☐ 144 sq km/yr

Water Use in Arid Regions

3. Water usage in arid regions is becoming an increasingly important issue. *[Note: This was Problem 2 in the Desert Landscapes worksheet, but is appropriate for this worksheet as well (especially if you did not do the Desert Landscapes worksheet)].*
(a) Check and double-click the placemarks for Problems 3a-i and 3a-ii to fly to the Phoenix, AZ area. How are the areas highlighted by the placemarks different?
☐ 3a-i: lush, green golf course and 3a-ii: arid desert with dry washes
☐ 3a-i: arid desert with dry washes and 3a-ii: lush, green golf course
(b) What are the features highlighted by the Problem 3b placemarks?
☐ landslides
☐ marinas
☐ dams
☐ wildlife sanctuaries
(c) What is the feature highlighted by placemark Problem 3c?
☐ interstate
☐ water canal
☐ railroad
☐ bike path
(d) Follow the feature in Problem 3c to the west (continue across the tunnels/pipelines that look like gaps). Where does it ultimately originate (the placemarks for Problem 3d mark the ends of a long tunnel)?
☐ Colorado River
☐ Pacific Ocean
☐ a glacier in the Sierra Nevada Mountains
☐ Gulf of California

Just for Fun... Considering your answers to the questions above and to their arid location, are Phoenix and the surrounding communities sustainable? Can they grow larger? Should they grow larger?

Preservation of Natural Habitats

4. Check and double-click placemark Problem 4 to fly to the Everglades in Florida. In the Layers panel turn on *More > Parks and Recreation Areas > US National Parks* to see the green outline of the boundary of the protected national park (note the sharpness of this eastern boundary with developed areas).
(a) Based on the alignment of hammocks of grass (the elongate ridges that create a grain to the Everglades), and on the position of the coast relative to this area, what is the direction of regional water flow in the Everglades?
☐ NNE to SSW
☐ NNW to SSE
(b) North of the park border, if developers or farmers drain and utilize this land, what will happen to the source of water for the Everglades, and therefore to the ecosystem of the Everglades?
☐ the Everglades will be unchanged
☐ the Everglades will become significantly drier

Developing Your Own Interactive Google Earth Materials

M. Scott Wilkerson & M. Beth Wilkerson

© 2011 Europa Technologies
US Dept of State Geographer
© 2011 Google
© 2011 MapLink/Tele Atlas

lat 37.529248° lon -94.621192° elev 257 m

Google earth

Eye alt 8391.75 km

In order for users to see and use personal media files (e.g., images, diagrams, maps, etc.) that you have embedded in your Google Earth project, you have to make the files publicly accessible from a web server rather than simply storing the files on your hard disk (otherwise users cannot see your images and usually will see only a box with a red "X" in it). There are many websites that will allow you to store data online at little to no cost (e.g., Dropbox, Flickr, Picasa, Facebook, etc.). Basically, all you need is a URL for your image so it can be viewed in a standard web browser. Below we describe how to set up and use Dropbox.

1. Using your web browser, go to the following website:
   http://www.dropbox.com

2. Click the **Log in** link in the upper righthand corner of the page.

3. Select **Create an Account** to create a free 2-GB account. A dialog box appears where you enter your **name**, **email address**, and **password**.

4. The Dropbox desktop application will automatically download. You can install this on your computer, or you can just use the Dropbox website (we'll continue using the website).

5. Click the **Back to home** link to go to your Dropbox home area. Feel free to explore the links on the **Get Started tab**. However, for now, click on the **Files tab**. You should see links to a Getting Started.pdf file, a Photos folder, and a Public folder with a series of buttons/links immediately above the list of files/folders. We're most interested in the Public folder since this is the only folder where you can share media with others.

6. Click on the **Public folder** to open that folder (feel free to read the How to use the Public folder.rtf file). You can click the **Upload button/link** to find and upload an image from your computer to the Public folder using your system's standard file dialog box.

7. To obtain the URL for this image, highlight the file with your mouse. Notice that as the mouse passes over the file, a **downward-pointing triangle** pops up on the **right side** of the window. Click the triangle, select **Copy public link**, and then click **Copy to clipboard**. This URL can then be pasted into Google Earth, a web browser, or any text editor to let others reference your file. You might also have noticed that the downward-pointing triangle also is home to the **Download file** command, among others.

8. To return back to your Home folder, click on the **Parent folder** link. This will move you back up the directory structure one folder at a time.

Dropbox has many other features and uses, but these are beyond the scope of this workbook. If you use one of the many other services, you will likely go through similar steps of setting up an account, uploading files, and then obtaining the URL reference address for those files. *Please also recall that you can reference images already on the web by **clicking RMB on the image** and selecting **Copy Image Address/Location**.* You should always first ask permission to reference someone else's images and realize that they may delete, move, or rename the images at any time, thus breaking your URL reference link to its location.

Module **1**

Placemarks

In this module, you will learn how to create and manipulate placemarks within Google Earth.

Project 1.1	creating placemarks

1. Type Mount Saint Helens in the **Fly to** text field of the **Search** panel. Press **Enter/Return** or click the **Magnifying Glass** icon to fly to this area. Placemarks available in Google Earth appear in the scrollable window below the Search text field (Figure 1.1). Clicking the **X** in the lower right corner of the Search panel (Figure 1.1) erases your search results.

Figure 1.1: Search panel for
Mount Saint Helens.

2. Click on the **Add Placemark** icon ⬚ on the toolbar (or **Add > Placemark** from the menu). A new placemark appears in the Google Earth viewer with a yellow box around the icon.

3. In the **New Placemark** dialog, type Mt. St. Helens in the **Name** text field.

4. You can move the new placemark icon in the Google Earth viewer by moving your cursor over the blinking yellow box (the cursor should change to a hand with a pointing finger). Click and drag the icon to its proper location. For this project, move the placemark just to the NNE of the volcano to a position on Spirit Lake. *Note that it is not uncommon for the **New Placemark** dialog to obscure the new placemark icon; if it does, click on the dialog's title bar and drag it to one side.*

97

5. When you're finished, click **OK**. That's it! You've made a placemark! You can now visit somewhere else in Google Earth and instantly return to this location by simply locating this placemark in your **Places** panel and double-clicking it.

Where is my placemark stored?

Once you have clicked OK in the New Placemark dialog, the new placemark is typically added somewhere in the Places panel. To direct the placemark to be added to a specific folder, select the folder before adding the placemark <u>or</u> RMB on the folder and select Add > Placemark from the menu that appears. To copy a placemark, select the object to be copied, click RMB, and select Copy from the menu that appears. Then, select a folder, click RMB, and select Paste from the menu that appears. To store your placemark permanently, you should make sure that it is saved somewhere in your My Places folder.

Would you like to make this placemark more informative? We're going to learn how to do that in the next project on editing placemarks.

Why is text entered into the Name text field sometimes not kept?

This happens not only for the Name text field, but also occasionally for other text fields as well (e.g., Links to images, movies, etc.). The easiest workaround for this issue is to either tab out of the text field or simply click in another element of the dialog box (e.g., the Description text field). Your entries should then be retained.

Project 1.2   editing placemarks

1. Click RMB on the "Mt. St. Helens" placemark that we created in Project 1.1 (either in the **Places** panel or in the Google Earth viewer). A context menu appears with several operations that you can perform on this placemark. For this project, select **Get Info** (Mac) or **Properties** (PC). The **Edit Placemark** dialog appears (Figure 1.2).

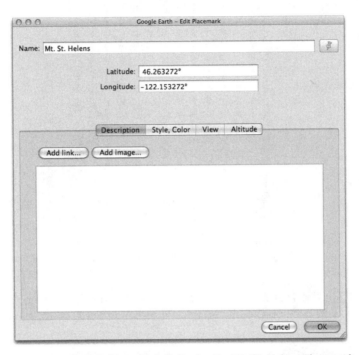

Figure 1.2: The Edit Placemark dialog for the Mt. St. Helens placemark.

2. You can edit the following features of your placemark:

- <u>Name</u>: Change the text in the **Name** text field to change the name of the placemark.

- <u>Latitude & Longitude</u>: These fields show the exact location of the placemark and will display the coordinates in the format specified in the **Preferences** (Mac) or **Options** (PC). In Figure 1.2, latitude and longitude are shown in decimal degrees.

  These numbers can be changed manually by entering values in the text fields, or they will automatically change as you move the placemark. *(Note: You must be in Edit mode, with a yellow box around the placemark, to move the placemark.)*

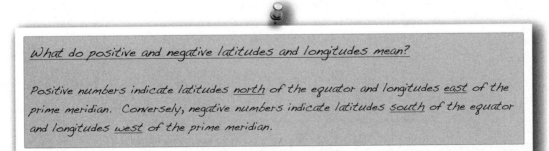

What do positive and negative latitudes and longitudes mean?

Positive numbers indicate latitudes <u>north</u> of the equator and longitudes <u>east</u> of the prime meridian. Conversely, negative numbers indicate latitudes <u>south</u> of the equator and longitudes <u>west</u> of the prime meridian.

- Icon: Click on the icon to the right of the **Name** text field to open the **Icon** dialog (Figure 1.3). Here you can select a different icon; use a custom icon; change the color, scale, or opacity of an icon; remove an icon; etc. Once you've made changes, click **OK**.

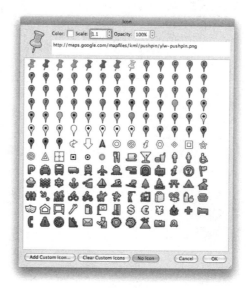

Figure 1.3: The Icon dialog allows considerable flexibility in modifying placemark icons.

Description tab: The **Description** field allows you to enter information that you want to pop up when you click on the placemark. For this project, enter the text in Figure 1.4.

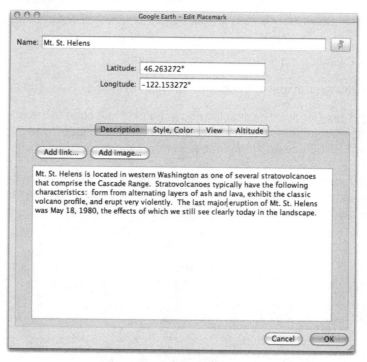

Figure 1.4: The Description field allows you to provide the user with information about your placemark.

100

- <u>Style, Color tab</u>: The **Style, Color** tab allows you to customize the color, size, and opacity of your placemark icon and its label. Please feel free to experiment with these settings on your own.

- <u>View tab</u>: The **View** settings allow you to customize the view that you see when you fly into a placemark. View information (latitude, longitude, range, heading, and tilt) may be entered manually, or you can set the view using the Google Earth viewer navigation controls and then click **Snapshot current view** to retain the zoom, perspective, and position. Zoom into the Spirit Lake area in the Google Earth viewer and tilt your perspective to the S such that you can see the topographic effects of the volcano (e.g., Figure 1.5) and then click **Snapshot current view** to change the settings (Figure 1.6).

Figure 1.5: A perspective view to the south of the Mount Saint Helens' volcanic cone.

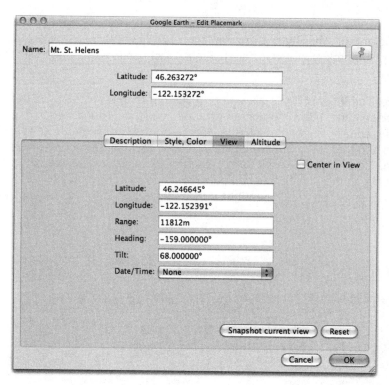

Figure 1.6: Settings in the View tab created by clicking
Snapshot current view for the view in Figure 1.5.

- Altitude tab: The **Altitude** tab specifies where the placemark icon is located relative to the map. Unless you want your icon floating above the Earth's surface, leave the altitude at 0m. Please feel free to experiment with this setting on your own.

- Add link... button: The **Add link...** button inserts the formatted code at the cursor's location in the **Description field** to create a clickable URL within your placemark (see Project 1.5).

- Add image... button: The **Add image...** button writes the formatted code at the cursor's location in the **Description field** for embedding an image within your placemark (see Project 1.4).

3. Click **OK** to apply your changes and to close the **Edit Placemark** dialog.

4. Now zoom out until you can see the entire planet, and then double-click on your "Mt. St. Helens" placemark. You should now fly in to your location and stop at the perspective and zoom level that you set. In addition, the descriptive text that you entered for the placemark now appears in a pop-up balloon.

Congratulations! You've just learned how to turn an ordinary placemark in Google Earth into an informative kiosk! Now, how about adding some pizazz to your placemark text by formatting it?

Project 1.3   basic formatting of placemark text

Placemark text can be formatted using HTML tags, the same code that is used to format web pages.

1. Click RMB on the "Mt. St. Helens" placemark that we edited in Project 1.2, and select **Get Info** (Mac) or **Properties** (PC) from the context menu. The **Edit Placemark** dialog appears.

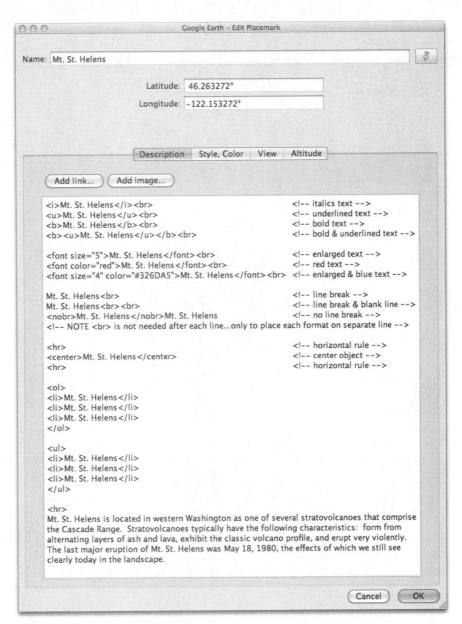

Figure 1.7: Basic HTML tags used to format placemark text.
Note that you can use some basic color names in place of the hexadecimal values.

2. Figure 1.7 shows the **Description** field with some basic formatting tags, and Figure 1.8 shows the resulting output.

103

What are formatting tags?

Formatting tags open and close around the text on which they act (and if you open something, you must close it). When several tags are nested (i.e., act on the same text), they must be closed in the reverse order in which they were opened. Common font colors can be specified by name, whereas other colors can be specified in hexadecimal format.

Mt. St. Helens

Mt. St. Helens
Mt. St. Helens
Mt. St. Helens
Mt. St. Helens
Mt. St. Helens
Mt. St. Helens
Mt. St. Helens
Mt. St. Helens
Mt. St. Helens

Mt. St. HelensMt. St. Helens

Mt. St. Helens

---

1. Mt. St. Helens
2. Mt. St. Helens
3. Mt. St. Helens

- Mt. St. Helens
- Mt. St. Helens
- Mt. St. Helens

Mt. St. Helens is located in western Washington as one of several stratovolcanoes that comprise the Cascade Range. Stratovolcanoes typically have the following characteristics: form from alternating layers of ash and lava, exhibit the classic volcano profile, and erupt very violently. The last major eruption of Mt. St. Helens was May 18, 1980, the effects of which we still see clearly today in the landscape.

Directions: To here - From here

Figure 1.8: Formatted placemark using the tags in Figure 1.7.

How can I determine hexadecimal color values?

There are many free resources [e.g., online hexadecimal color charts, freeware (PC), & Digital Color Meter (Mac)] as well as commercial software (e.g., Adobe Photoshop).

3. Let's edit the "Mt. St. Helens" placemark again, and this time include the following in the **Description** text field:

```
<hr>
<center>
<font size="5" color="#4E7646">Mt. St. Helens</font><br>
<i>Cascade Range, western Washington</i>
</center>
<hr>
<b>Mt. St. Helens</b> is located in western Washington as one of several <i>stratovolcanoes</i>
that comprise the Cascade Range.  <i><u>Stratovolcanoes</u></i> typically have the following
characteristics:
<ol>
<li>form from alternating layers of ash and lava,</li>
<li>exhibit the classic volcano profile, and </li>
<li>erupt very violently.</li>
</ol>
<br>
The last major eruption of Mt. St. Helens was <b>May 18, 1980</b>, the effects of which we still see
clearly today in the landscape.
<hr>
```

4. Figure 1.9 shows the resulting formatted placemark from the text in step 3.

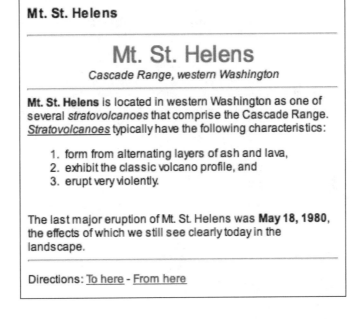

Figure 1.9: Formatted placemark using the tags in Step 3.

5. Great! You now have a basic understanding of how to format placemark text...but there is still more that you can do with placemarks! Before we go on, now would be a good time to practice some of the concepts that you've learned.

Project 1.4 | inserting images into placemarks

Images within placemarks are an important way to convey additional information about a location, whether it be a graph of data, a photo, or a map. To insert an image into a placemark, you must first have its Internet URL address (that is, the image must be located on a web server). For most images on the Web, you can simply click RMB on the image in a web browser and copy its URL *(just realize that the availability of that image depends entirely on that website and web server)*.

1. In a web browser, navigate to the USGS/Cascades Volcano Observatory website (http://vulcan.wr.usgs.gov/Volcanoes/MSH/framework.html). Locate a photo of Mount Saint Helens (http://vulcan.wr.usgs.gov/Imgs/Jpg/MSH/Images/MSH82_st_helens_plume_from_harrys_ridge_05-19-82_med.jpg), and copy the URL address from the browser URL field.

2. Click RMB on the "Mount Saint Helens" placemark that we edited in Project 1.3, and select **Get Info** (Mac) or **Properties** (PC) from the context menu. The **Edit Placemark** dialog appears.

3. In the **Description** text field, add "<center></center>" tags, and place your cursor between the tags. Click the **Add image...** button, paste the URL text for your image in the URL text box that appears, and click OK to the right of the **Image URL** text box. Google Earth will insert the properly formatted <img src=" "> tags into your **Description** text field to create a centered image in your placemark (Figure 1.10). *Note: If you omit the <center> tags, the image will be left-justified.* To change the image, simply change the URL in src="URL".

> <img src="http://vulcan.wr.usgs.gov/Imgs/Jpg/MSH/Images/MSH82_st_helens_plume_from_harrys_ridge_05-19-82_med.jpg"/>

Congratulations! Inserting images really just entails adding this one line of code to your **Description** text field (multiple images can be added as well). The real work with images comes up front when you must reduce the size of the images while retaining reasonable quality such that the images are small enough to download quickly.

Can I use images from my hard disk?

Yes...but you will be able to see the image only from that computer. Consider storing your images on a web server like Dropbox (see Before You Begin at the start of Section 3) if you want others to have access to your images.

PC: <img src="C:\Documents and Settings\Username\MyPictures\Image.jpg">
where C:\Documents and Settings\Username\MyPictures\ is the path to the folder containing the jpeg image, Image.jpg.

Mac: <img src="/Users/Username/Pictures/Image.jpg">
where /Users/Username/Pictures/ is the path to the folder containing the jpeg image, Image.jpg.

Mt. St. Helens

Mt. St. Helens
Cascade Range, western Washington

Mt. St. Helens is located in western Washington as one of several *stratovolcanoes* that comprise the Cascade Range. *Stratovolcanoes* typically have the following characteristics:

1. form from alternating layers of ash and lava,
2. exhibit the classic volcano profile, and
3. erupt very violently.

The last major eruption of Mt. St. Helens was **May 18, 1980**, the effects of which we still see clearly today in the landscape.

Directions: To here - From here

Figure 1.10: Formatted placemark with a centered image.

Project 1.5 inserting URL links into placemarks

Sometimes you want to link to additional Internet content outside of Google Earth. In such instances, it would be nice to have a clickable link that opens a web browser to the appropriate page. For example, we would be remiss in not referencing the image that we obtained from the USGS/Cascades Volcano Observatory website. Here's how we do that.

1. In a web browser, navigate to the USGS/Cascades Volcano Observatory website and locate the photo of Mt. Saint Helens that we used (http://vulcan.wr.usgs.gov/Imgs/Jpg/MSH/Images/MSH82_st_helens_plume_from_harrys_ridge_05-19-82_med.jpg). Copy the URL address from the browser URL field.

2. Click RMB on the "Mt. Saint Helens" placemark that we edited in Project 1.4, and select **Get Info** (Mac) or **Properties** (PC) from the menu. The **Edit Placemark** dialog appears.

3. In the **Description** text field, add "
 Image from USGS/Cascades Volcano Observatory website." after your image, but before the </center> tag, to center your URL link on the line beneath the image. Select/highlight "USGS/Cascades Volcano Observatory", click the **Add link...** button, paste the URL location for your image in the URL text box that appears, and click OK next to the **Link URL** text box. Google Earth will insert the properly formatted <a href=" "> tags into your **Description** text field to create a clickable URL in your placemark (Figure 1.11). *Note: If you don't select text to become the link, Google Earth will insert the actual URL as the clickable text for your link.*

>
Image from <a href="http://vulcan.wr.usgs.gov/Imgs/Jpg/MSH/Images/MSH82_st_helens_plume_from_harrys_ridge_05-19-82_med.jpg">USGS/Cascades Volcano Observatory</a> website.</center>

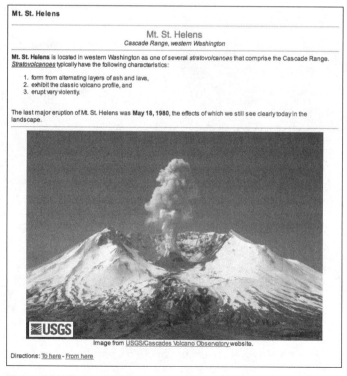

Figure 1.11: Formatted placemark with a centered image & URL link.

108

Project 1.6   inserting YouTube video into placemarks

Video is yet another powerful means of conveying information about a place. Google Earth has the capability to use "embed" code within a placemark to link to web videos that offer that option. If there is not a direct embed link and if the web video is based on Adobe Flash, you might be lucky enough to copy the Flash embed code directly by clicking RMB on the web video and pasting it into a Google Earth placemark.

For this project, we are going to provide instructions on using web videos from the popular YouTube website. It is slightly more involved than what you learned about inserting images into placemarks in Project 1.4 because you have to obtain the embed code; however, it still is fairly straightforward.

1. In a web browser, go directly to a USGS video about Mt. St. Helens by typing the following URL (http://youtu.be/sC9JnuDuBsU).

2. Click on the **Share** button beneath the video. An information frame expands. Click the **Embed** button and copy the highlighted code that appears in the expanded frame.

3. In Google Earth, click RMB on the "Mt. Saint Helens" placemark that we edited in Project 1.3 (or Projects 1.4–1.5, although those will have the image that was inserted into the placemark) and select **Get Info** (Mac) or **Properties** (PC) from the context menu. The **Edit Placemark** dialog appears.

Can I embed my own movies into Google Earth placemarks?

Yes...however, it is a non-trivial process beyond the scope of this workbook. The most straightforward approach is to upload your video to YouTube and then use the technique described in Project 1.6.

4. In the **Description** text field, type the bold tags and paste the YouTube text between them at the bottom of your description (the Mount St. Helens: A Catalyst for Change | USGS YouTube Channel is only to reference the source). Give the video a few seconds to load as it is over 6 minutes long. *(Note: If you omit the <center> tags, the video will be left-justified.)*

```
<center>
<iframe width="420" height="315" src="http://www.youtube.com/embed/
sC9JnuDuBsU" frameborder="0" allowfullscreen></iframe>
<br>
Mount St. Helens: A Catalyst for Change | USGS YouTube Channel
</center>
<br><br>
```

5. Congratulations! You now have an enormous source of YouTube video material at your disposal to add to your placemarks (see Figure 1.12)!

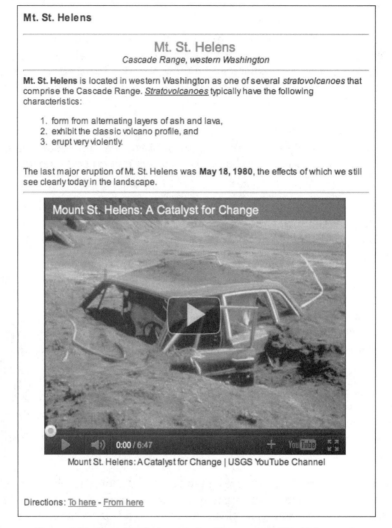

Mt. St. Helens

Mt. St. Helens
Cascade Range, western Washington

Mt. St. Helens is located in western Washington as one of several *stratovolcanoes* that comprise the Cascade Range. *Stratovolcanoes* typically have the following characteristics:

1. form from alternating layers of ash and lava,
2. exhibit the classic volcano profile, and
3. erupt very violently.

The last major eruption of Mt. St. Helens was **May 18, 1980**, the effects of which we still see clearly today in the landscape.

Mount St. Helens: A Catalyst for Change

0:00 / 6:47

Mount St. Helens: A Catalyst for Change | USGS YouTube Channel

Directions: To here - From here

Figure 1.12: Formatted placemark with a centered YouTube movie.

110

Module **2**

Paths & Polygons

This chapter will introduce you to how you can use paths and polygons in Google Earth.

Project 2.1 creating paths

1. For this project, we're going to first fly to Pompei, Italy. Type Pompei, Italy in the **Fly to** text field in the **Search** panel, and click the **Magnifying Glass** icon to fly there.

2. Add a placemark named "Pompeii" and move it about 1000 m west of where Google marked the modern city of Pompei, Italy (yes, they are spelled differently!). Set an oblique perspective to the NW with an eye altitude of 775 m (Figure 2.1). This new placemark, as most archeology buffs know, marks the ancient Roman city of Pompeii, which was partially buried in 79 C.E. by a pyroclastic eruption of the nearby volcano Mt. Vesuvius. A great deal of the city has now been excavated to provide a wonderful glimpse into everyday Roman culture and society.

Figure 2.1: Perspective view of the ancient Roman city of Pompeii with Mt. Vesuvius in the background.

3. Navigate to a view where you are looking vertically down on Mt. Vesuvius and Pompeii. We are going to create a path from the volcano to the city of Pompeii.

111

4. To create a new path, click the **Add Path** icon  or click RMB in the **Places** panel and choose **Add > Path**. The **New Path** dialog appears, and your cursor changes to a crosshair in order to digitize points on the path. Name the path "Pyroclastic Flow".

5. Begin by clicking a point within Mt. Vesuvius's crater (you may have to move the **New Path** dialog). Then click a few more points to create a white path across the landscape from Mt. Vesuvius to Pompeii (Figure 2.2). Click **OK** to finish constructing the path.

Figure 2.2: Vertical view of the path constructed from Mt. Vesuvius to the ancient city of Pompeii.

Very good! In Module 4, we'll learn how to automatically fly a tour along this path to actually simulate the pyroclastic flow that destroyed Vesuvius.

Project 2.2   editing paths

1. Click RMB on the "Pyroclastic Flow" path that we created in Project 2.1 (either in the **Places** panel or in the Google Earth viewer). A context menu appears with several operations that we can perform on this path. For this project, select **Get Info** (Mac) or **Properties** (PC). The **Edit Path** dialog appears (Figure 2.3).

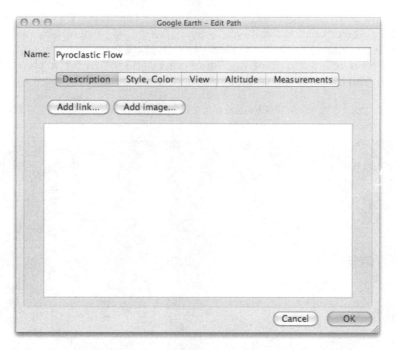

Figure 2.3: The Edit Path dialog for the Pyroclastic Flow path.

2. You can edit the following features of your path (many of them are similar to editing a placemark):

- Name: Change the text in the **Name** text field to change the name of the path.

- Description tab: The **Description** field allows you to enter information that you want to pop up when you click on the path. For this project, we'll not include any text here.

- Style, Color tab: The **Style, Color** tab allows you to customize the color, width, and opacity of the path line. For this project, change the color of the line to blue and increase its width to 2.0.

- View tab: The **View** tab really isn't all that beneficial for editing paths. The **Snapshot current view** function controls the view when you double-click the path.

- Altitude tab: The **Altitude** tab specifies where the path line is located relative to various surfaces associated with the map. For this project, select **Relative to ground** from the drop-down menu and enter "500" into the **Altitude** text field. Next, check the **Extend path to ground** checkbox to create a skirt beneath the line that extends to the ground (when this feature is active, you can change its style and color in the **Style, Color** tab).

113

- Measurements tab: The **Measurements** tab provides the length of the path in the specified units.

- Add link... button: The **Add link...** button inserts the formatted code at the cursor's location in the **Description** field to create a clickable URL within your path balloon (see Project 1.5).

- Add image... button: The **Add image...** button writes the formatted code at the cursor's location in the **Description** field for embedding an image within your path balloon (see Project 1.4).

- Last, and perhaps most important, the individual path points can be edited (you may need to move the **Edit Path** dialog out of your way). That is, the points that you digitized to create the path become highlighted in red when the **Edit Path** dialog is active. If you pass the cursor over any red point, it will turn green, and the cursor will turn into a hand with a finger. You then can adjust that point on the path.

3. Click **OK** to apply your changes and to close the **Edit Path** dialog. Figure 2.4 shows your results.

Figure 2.4: The Pyroclastic Flow path with some of its characteristics modified using the Edit Path dialog.

What is a Smoot?

Oliver Smoot '62 was the shortest pledge in the Lambda Chi Alpha fraternity at MIT in Cambridge, MA. In 1958, his fraternity brothers decided to use his 5-foot, 7-inch (~1.70 m) frame as a measuring stick for the Harvard Bridge (which was 364.4 Smoots long). This "unit" of measurement has endured and has been included in Google Earth. (http://en.wikipedia.org/wiki/Smoot and references therein).

Project 2.3   measuring lines & paths

1. For this project, we're going to first fly to the South Rim of the Grand Canyon. Type the coordinates 36.057015, -112.138989 in the **Fly to** text field in the **Search** panel, and click the **Magnifying Glass** icon to fly there.

2. Click on the **Ruler** tool [icon] and select either **Line** or **Path**.

   - The **Line** tab allows a line to be drawn between two endpoints and displays the **length** in the units in the drop-down menu and the **heading** in degrees (Figure 2.5). **Map Length** provides the horizontal distance between the endpoints, whereas **Ground Length** approximates the ground distance by taking into consideration the elevation and map distance of the endpoints. *Step 4 (Show Elevation Profile) provides a more accurate measurement of the true ground distance.*

   - The **Path** tab permits you to draw a path comprised of multiple points and to show the **cumulative length** in the units specified in the drop-down menu.

   - The Line or Path can either be removed by clicking the **Clear** button or added as a new path using the **Save** button (in which case the **New Path** dialog appears).

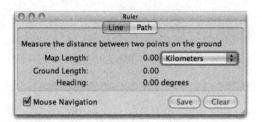

*Figure 2.5: Ruler dialog with
the Line tab selected.*

3. Use the **Path** tab in the **Ruler** to create, measure, and save a path between visitor centers on the North and South Rims (Figure 2.6; Map=17.29 km and Ground=17.30 km). You may want to turn on *More > Parks/Recreation Areas > US National Parks > Park Descriptions* in the **Layers** panel to see the visitor centers.

4. Now click RMB on the "visitor centers" path that we just saved (either in the **Places** panel or in the Google Earth viewer). From the context menu that appears, you have two additional ways to obtain measurements for the path:

   - Select **Get Info** (Mac) or **Properties** (PC) from the context menu, and the **Edit Path** dialog appears (Figure 2.3). Select the **Measurements** tab to display the distance between the visitor centers (this is equivalent to the **Ground Length** displayed by the **Ruler** tool).

   - Select **Show Elevation Profile** from the context menu, and a new window that shows the elevation changes along the path appears at the bottom of your screen (Figure 2.7). As you move your mouse within the **Elevation Profile** window, a red arrow in the Google Earth window shows the corresponding point along your path. The display also shows the slope of the ground surface as well as the **actual terrain distance** along the path (*for a profile to show sea-floor bathymetry for paths over oceans, select **Altitude > Clamp to sea floor** in the **Edit Path** dialog box*). Note that the terrain distance displayed in the

Elevation Profile window is greater than the distances displayed in the **Edit Path** or **Ruler** dialogs because detailed topography is taken into account (21.6 km for the visitor centers' path). The vertical exaggeration and detail of the profile cannot be changed. Click the "X" on the right side of the **Elevation Profile** window to close it.

Figure 2.6: A path drawn using the Ruler tool and then saved as a path.

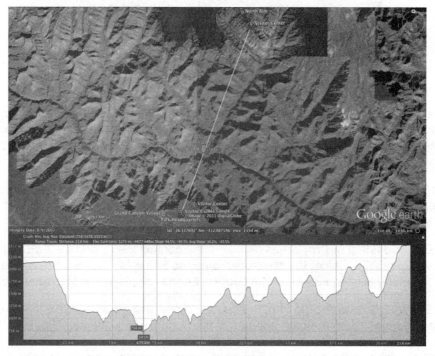

Figure 2.7: With Show Elevation Profile selected, users visualize the terrain and measure slope and actual terrain distances along a specified path.

Project 2.4	creating polygons

1. Type coordinates 35.027203, -111.022724 in the Fly to text field in the **Search** panel, and click the **Magnifying Glass** icon to fly to Meteor Crater (aka Barringer Crater).

2. Navigate to a view where you are looking vertically down on the crater (Figure 2.8). We are going to create a polygon of the crater rim in order to estimate the area of the crater.

3. To create a new polygon, click the **Add Polygon** icon  or right-click in the **Places** panel and choose **Add > Polygon**. The **New Polygon** dialog appears, and your cursor changes to a crosshair in order to digitize points on the polygon. Name the polygon "Meteor Crater". As you create the polygon, portions of the image may be obscured. If so, click on the **Style, Color** tab and set the **Area Opacity to 50%**.

4. Begin by clicking a point along the crater rim (you may have to move the **New Polygon** dialog). Then click additional points to create a white polygon that outlines the crater rim (Figure 2.8). Click **OK** when you finish constructing the polygon.

Figure 2.8: Vertical view of Meteor Crater during the initial stages of constructing a polygon to outline the crater rim.

Great! In the next project, we'll learn how to edit the polygon to change the points and to change the color of the polygon.

Project 2.5   editing polygons

1. Click RMB on the "Meteor Crater" polygon that we created in Project 2.4 (either in the **Places** panel or in the Google Earth viewer). A context menu appears with several operations that we can perform on this polygon. For this project, select **Get Info** (Mac) or **Properties** (PC). The **Edit Polygon** dialog appears (Figure 2.9).

Figure 2.9: The Edit Polygon dialog for the Meteor Crater polygon.

2. You can edit the following features of your polygon (many of which are similar to editing a path):

   • Name: Change the text in the **Name** text field to change the name of the polygon.

   • Description tab: The **Description** field allows you to enter information that you want to pop up when you click on the polygon. For this project, we'll not include any text here.

   • Style, Color tab: The **Style, Color** tab allows you to customize the color, width, and opacity of the polygon outline and fill (area). For this project, change the color of the line to black and increase its width to **2.0**. Change the color of the area to blue and make the area have a **50% opacity**.

   • View tab: The **View** tab really isn't all that beneficial for editing polygons. The **Snapshot current view** function controls the view when you double-click the polygon.

   • Altitude tab: The **Altitude** tab specifies where the polygon is located relative to various surfaces associated with the map. For this project, leave it as **Clamped to ground**.

- Add link... button: The **Add link...** button inserts the formatted code at the cursor's location in the **Description** field to create a clickable URL within your polygon balloon (see Project 1.5).

- Add image... button: The **Add image...** button writes the formatted code at the cursor's location in the **Description** field for embedding an image within your polygon balloon (see Project 1.4).

- Lastly, just as for paths, the individual points that comprise the polygon can be edited (you may need to move the **Edit Polygon** dialog out of your way). That is, the points that you digitized to create the polygon become highlighted in red when the **Edit Polygon** dialog is active (Figure 2.9). If you pass the cursor over any red point, it will turn green, and the cursor will turn into a hand with a finger. You then can adjust that point on the polygon outline. Additional points can be added by clicking on the line next to the active (blue) dot.

3. Click **OK** to apply your changes and to close the **Edit Polygon** dialog.

How can I measure polygon regions?

Direct area measurement of polygons has been reserved for Google Earth Pro. For features that have regular geometric shapes, areas can be calculated using the Ruler tool (map distances) and the Elevation Profile tool (terrain distances). Linear distances (e.g., width, length, diameter, perimeter, etc.) also can be measured using these tools. For example, the diameter of Meteor Crater is approximately 1.1–1.3 km.

119

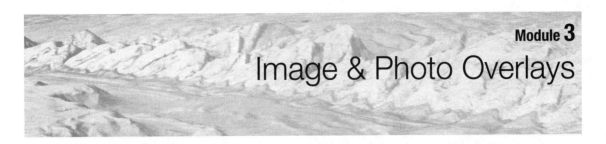

Module **3**

Image & Photo Overlays

Image and zoomable photo overlays are powerful features within Google Earth. This module focuses on how to georeference image overlays to their proper positions on the globe and how to create zoomable photo overlays within the Google Earth viewer.

Project 3.1   creating an image overlay

1. For this project, we're going to first fly to the Yucatán Peninsula. Type coordinates 20.188160, -88.859868 in the **Fly to** text field in the **Search** panel, and click the **Magnifying Glass** icon to fly there. Stop your descent at an altitude of ~600,000 m (~370 mi).

2. In the **Layers** panel, click **Borders and Labels** to turn on the region outlines.

3. To create a new image overlay, click the **Add Image Overlay** icon or click RMB in the **Places** panel and choose **Add > Image Overlay**. The **New Image Overlay** dialog and a green frame for your image appears (Figure 3.1).

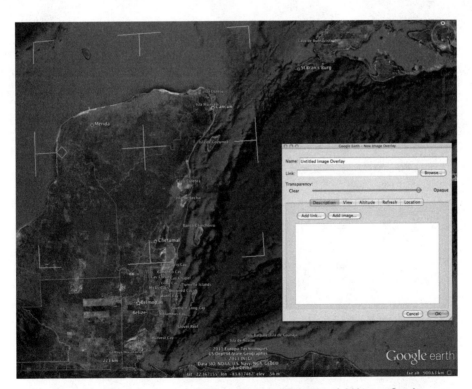

Figure 3.1: Green frame and dialog appear after selecting Add Image Overlay. The view has been moved back to ~900,000 m.

121

4. Images must be linked to the green frame in order to be displayed (see the **Link** text field in Figure 3.1). Such links can be created one of two ways:

   a. **From your local hard disk**: Click the **Browse** button and a standard **Open File** dialog for your computer system will appear, allowing you to link to an image (e.g., jpg, gif, tiff, png, etc.) on your computer. *The only issue with this approach is that the image overlay will only be viewable from your computer.* That is, if you save the KML/KMZ file and move it to another computer, you'll only see a box with a large "X" instead of your image overlay (you'll also see the "X" if you move your image file and don't update the Google Earth link).

   b. **From the Web**: Type or paste a URL for an image that is on a web server in the **Link** text field. For example, you can click RMB on most images on web pages, select **Copy Image Address/Location**, and paste it in the **Link** text field (of course, you should first request permission to use the image). Alternatively, for your personal images, you will need to load them to a web server where they can be referenced by a URL (see **Before You Begin** at the beginning of **Section 3** to learn about how to use Dropbox, or contact your technical support person about setting up a local web server).

5. In the **New Image Overlay** dialog, change the **Name** to "Satellite Overlay" and insert "http://media.wwnorton.com/college/geo/geotours/ChicxulubSatImage.jpg". Click **OK**.

That is all there is to it! You'll notice, however, that the image overlay is not scaled and aligned to exactly match the yellow country borders. In Project 3.2, we'll learn how to edit the image overlay to georeference it with the Google Earth virtual globe.

Project 3.2   editing an image overlay

1. Click RMB on the "Satellite Overlay" overlay that we created in Project 3.1 (in the **Places** panel). For this project, select **Get Info** (Mac) or **Properties** (PC). The **Edit Image Overlay** dialog appears (Figure 3.2).

Figure 3.2: The Edit Image Overlay dialog with the green frame containing the non-georeferenced image overlay. Satellite image courtesy of NASA.

2. The editing objective is to manually adjust the image overlay to where it matches (as best you can) the area that it represents. In this example, we'll georeference the image by aligning the shoreline of the image with the yellow border outlining the Yucatán Peninsula. You can facilitate this process by moving the **Transparency** slider in the **Edit Image Overlay** dialog to a semi-transparent setting in order to better see how to align the image (you may want to move the **Edit Image Overlay** dialog out of the way).

3. To align the image, you will manipulate the green frame. To do so, move the cursor over the various frame features, and the cursor will change from the panning cursor (an open hand) to the editing cursor (a hand with a pointing finger) when it is over an editable part of the frame:

   - **Corners**: Changing the position of each corner will scale the frame by moving the two adjacent sides. If you hold down the Shift key, the entire frame is scaled proportionally.

   - **Sides**: Changing the position of a side will scale the frame by moving only that side. If you hold down the Shift key, the entire frame is scaled proportionally.

123

- **Center Cross**: Changing the position of the center cross will move/translate the entire frame.

- **Diamond** (left side of frame): Clicking and dragging within the diamond will rotate the frame.

4. Figure 3.3 shows a georeferenced image overlay that has been adjusted "by eyeball" to obtain a reasonable fit (yours should look something like this). Please note that some images (e.g., maps, aerial photos, etc.) may not fit perfectly because of their projection or because of drafting errors in generating the map. Before you select **OK** to accept your adjustments, you probably will want to move the **Transparency** slider back to **Opaque**.

Figure 3.3: A semi-transparent overlay adjusted to be georeferenced to the Google Earth virtual globe country borders.

5. In the **Places** panel, select "Satellite Overlay," and click the **Adjust Opacity** icon at the bottom of the **Places** panel. Moving the slider that appears to the left makes the image overlay more transparent. You can adjust your viewing position and the transparency to highlight features on the ground surface related to your image overlay.

Congratulations! You have just learned a very important process you can use to integrate all types of maps, diagrams, and photos with the Google Earth imagery—and see them in their proper spatial perspective. Project 3.3 will show you how to take this information and use it in a meaningful way. Before we conclude, however, please note that Google Earth Pro can automatically import georeferenced GEOTIFF images and shapefiles.

Project 3.3   measuring objects on an image overlay

1. Turn on the "Satellite Overlay" overlay that we georeferenced in Project 3.2 (click the checkbox in the **Places** panel). Note that the NW corner of the Yucatán Peninsula shows a semi-circular feature. This feature is a partial expression of the Chicxulub (CHICKS-oo-loob) Crater that formed when an asteroid hit this region approximately 65.5 million years ago, leading to the extinction of many species (most famously, the dinosaurs).

2. To further constrain the geometry of the crater, we'll create an image overlay that depicts the gravity signature of this area and the nearby offshore. Using the techniques described in Projects 3.1–3.2, create and georeference a gravity map of the region found at (http://media.wwnorton.com/college/geo/geotours/ChicxulubGravityMap.gif). The resulting image overlay is shown in Figure 3.4.

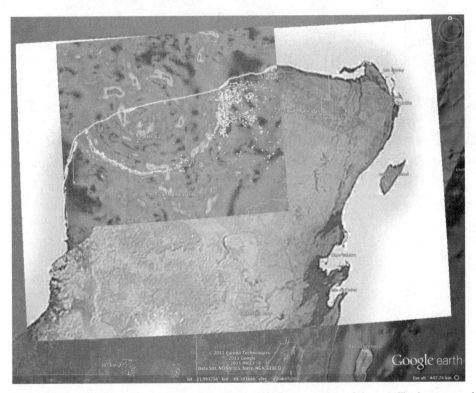

Figure 3.4: A gravity map image overlay of the Chicxulub Crater on top of the satellite image overlay of the area created and georeferenced in Projects 3.1–3.2. Gravity map courtesy of NASA.

3. Using the bull's-eye pattern on the gravity map and the maximum extent of the crater as depicted on the satellite imagery overlay, you can now add a path (Project 2.1) to measure the crater's diameter and/or add a polygon (Project 2.4) to represent the crater's area (Figure 3.5). *If the polygon is not displayed on top of the overlays, edit the polygon by going to the Altitude tab, selecting Absolute, and entering a value of 500 m.*

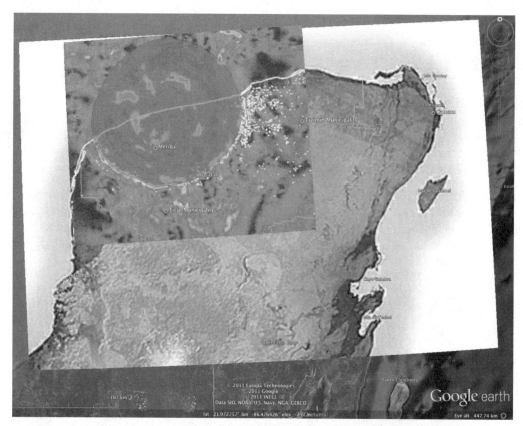

Figure 3.5: A semi-transparent polygon representing the Chicxulub Crater has a measured diameter of ~150 km (compare this with the diameter of Meteor Crater in Project 2.5).

Project 3.4 | **creating a zoomable photo overlay**

1. For this project, we're going to first fly to Yosemite National Park. Type the coordinates 37.746773, -119.591832 in the **Fly to** text field in the **Search** panel and click the **Magnifying Glass** icon to fly there. Rotate your perspective until you are viewing Half Dome toward the NE, as shown in Figure 3.6.

Figure 3.6: Oblique perspective view of Half Dome from the Yosemite Valley floor.
Vertical exaggeration of the terrain should be 1.

Half Dome is an exfoliation dome comprised of granite/granodiorite of the Sierra Nevada batholith. It is rounded because the igneous rock exfoliates in concentric layers much like peeling the layers of an onion. The name *Half Dome* originated because the flat NW face makes it appear as though half of the dome is missing. This face is one of the systematic fractures that exists in the area.

We're going to create a zoomable photo overlay to better visualize this amazing landform.

2. To create a new photo overlay, click RMB in the **Places** panel and choose **Add > Photo** (or select **Add > Photo** from the menu bar). The **New Photo Overlay** dialog appears (Figure 3.7).

3. Change the name to "Half Dome" and enter the URL of an image in the **Link** text field (http://media.wwnorton.com/college/geo/geotours/HalfDome.jpg; Figure 3.7).

Figure 3.7: Photo overlay of Half Dome.

4. Click **OK** to accept the photo overlay.

5. If you double-click on the Half Dome photo overlay in the **Places** panel, you will zoom into the photo, and the controls in the upper right corner of the Google Earth viewer will change. You can zoom/unzoom using SCROLL, or you can use the +/- controls in the upper right corner. The thumbnail in the upper right corner will develop a white outline as you zoom in on the photo. You can click and drag the white box to move your viewing location, or you can use the panning cursor (open hand). Click the **Exit Photo** button to exit the Zoomable Photo Overlay mode and to return to the Google Earth viewer. See Figure 3.8 to view the zoomable photo overlay controls.

128

*Figure 3.8: Double-clicking a photo overlay in the Places panel
allows the user to zoom into the photo at high resolution.*

Project 3.5 **editing a zoomable photo overlay**

1. Click RMB on the "Half Dome" photo overlay that we created in Project 3.4 (in the **Places** panel). Select **Get Info** (Mac) or **Properties** (PC). The **Edit Photo Overlay** dialog appears (Figure 3.9).

Figure 3.9: Clicking RMB on a photo overlay in the Places panel allows the user to edit the photo overlay.

2. You can change the positioning and perspective of the photo overlay using the controls in the **Photo** tab (Figure 3.9). This tool is very useful when aligning the photo with the Google Earth imagery. In Figure 3.9, the photo overlay was made semi-transparent to help with the alignment. Because the photo was taken from the ground, **Altitude** of the overlay was set to 2 m. The **Heading** (left-right), **Tilt** (up-down), and **Roll** (rotation) controls help align the position of the view behind the image by changing the camera view, whereas the **Horizontal** and **Vertical** controls help scale the view to the image. The vertical exaggeration in Google Earth should be set to 1 in order to minimize distortion *(Google Earth > Preferences (Mac) or Tools > Options (PC))*. Although aligning photo overlays is a trial-and-error process that requires some patience (especially if you don't know the exact location where the photo was taken), it can be a very powerful tool for facilitating detailed analysis of a feature in Google Earth (e.g., creating virtual "outcrops" for more detailed study of an area).

Module 4

Tours & Animations of Historical Imagery

Google Earth has several tools that allow you to take automated, prescribed tours and to view historical imagery for an area. This module describes how to use these tools.

Project 4.1	**creating tours of a folder of placemarks**

Google Earth allows you to select a folder in your **Places** panel and take an automated flying tour from placemark to placemark.

1. You can click once on any folder in your **Places** panel that contains placemarks to select it. For this project, we're going to use the "Project 4.1_Placemarks from Chap 10" folder in the **4. Tours & Historical Animations** folder (see **Places** panel in Figure 4.1). This folder of placemarks was created from John Wesley Powell's book, *The Exploration of the Colorado River and Its Canyons* (1997, Penguin Books), which chronicles the 1869 journey of Powell and his men down the unexplored and unmapped Colorado River in three wooden boats. Powell's book was derived from his journal, in which he provided dated entries keyed to prominent features along the river.

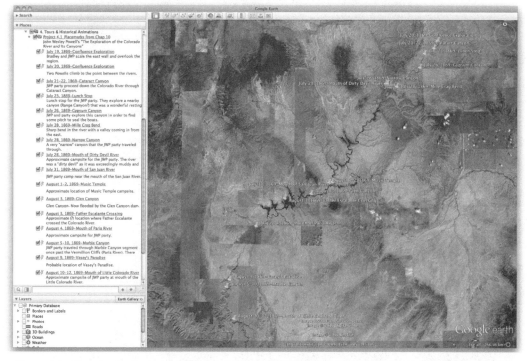

Figure 4.1: Placemarks from John Wesley Powell's book, The Exploration of the Colorado River and Its Canyons, document part of his 1869 journey down the unexplored and uncharted Colorado River.

If you want to create your own folder of placemarks, click RMB in the **Places** panel and select **Add > Folder**. Name the folder, then select it with a single mouse click. Any new placemarks created will go into this selected folder as long as it is highlighted. Alternatively, you can either drag or copy existing placemarks to the folder as well.

2. After you've highlighted the folder of placemarks, click on the **folder icon with the small triangle** (Figure 4.1; the triangle looks like a DVD player's Play button) in the lower right corner of the **Places** panel.

3. The **Tour Control** (Figure 4.2) will appear in the lower left corner of the Google Earth viewer window. The tour will begin with the first placemark and then will move sequentially to other placemarks.

Can I control the tour view?

Absolutely! You can edit each placemark's view as described in Module 1 to set the perspective, altitude, etc. You also can control aspects of the tour by using Google Earth > Preferences (Mac)/Tools > Options (PC) and then changing settings in the first section of the Touring tab. One neat feature is that you can automatically show your placemark balloons during the tour and specify their duration.

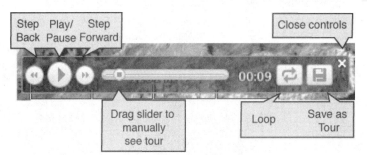

Figure 4.2: Tour Control.

4. The tour can be paused, looped continuously, or saved as a separate Google Earth tour in which you no longer need the placemarks (click the **Disk** icon in the **Tour Control** after playing the tour, see Project 4.2). If the tour isn't set to continuously loop, then it will stop and you can close the Tour Control by clicking the X. *Note: During the tour, the controls will become hidden so as to not obscure the view. To make them return, just move your mouse in the Google Earth viewer window.*

How can I make the placemarks invisible for tours?

It actually is fairly easy. Click RMB on the placemark in the Places panel and select Get Info (Mac) or Properties (PC). Click on the icon to change the icon to " No Icon" and name the placemark " " (an empty space). Alternatively, just play the tour and then save it using the Disk icon in the Tour Control, and the placemarks are no longer needed (see Project 4.2)

Project 4.2   creating tours along a specified path

Google Earth also offers the opportunity to fly tours along a prescribed path from the **Places** panel.

1. For this project, we're going to use the path that we created in Module 2 for the Roman city of Pompeii, Italy, that was partially buried by the eruption of Mt. Vesuvius. To fly to the region, double-click the "Pyroclastic Flow" path that you created. It is probably useful to check both the "Pyroclastic Flow" path and the "Pompeii" placemark so that they appear in the Google Earth viewer (Figure 4.3).

Figure 4.3: Vertical view of the path from Mt. Vesuvius to the ancient city of Pompeii.

2. We now want to fly along this path from the volcano to the city of Pompeii to simulate the 79 C.E. pyroclastic flow eruption of nearby Mt. Vesuvius that engulfed Pompeii. First, click the "Pyroclastic Flow" path in the **Places** panel once to highlight it.

3. The **Path Tour** icon  will appear at the lower right corner of the **Places** panel. Click on the **Path Tour** icon to automatically fly along the path from the first to the last point.

4. After taking an initial tour, you probably will want to adjust the settings. To do so, select **Google Earth > Preferences** (Mac) or **Tools > Options** (PC) and select the **Touring** tab (Figure 4.4).

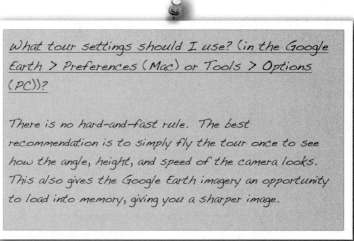

What tour settings should I use? (in the Google Earth > Preferences (Mac) or Tools > Options (PC))?

There is no hard-and-fast rule. The best recommendation is to simply fly the tour once to see how the angle, height, and speed of the camera looks. This also gives the Google Earth imagery an opportunity to load into memory, giving you a sharper image.

5. For this project, set the **Camera Tilt Angle** = 65, **Camera Range** = 100, and **Speed** = 900 (Figure 4.4).

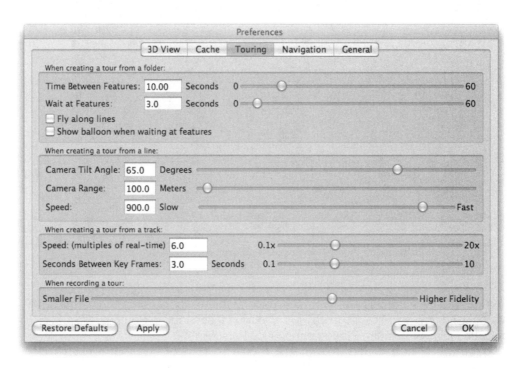

Figure 4.4: The Touring tab in the Preferences dialog. The middle section controls tours of paths.

6. Now play the tour again (but leave the "Pyroclastic Flow" path unchecked so that it is not visible in the Google Earth viewer). Note that the **Tour Control** appears (Figure 4.5).

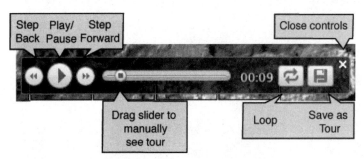

Figure 4.5: Tour Control.

7. After the tour has played, you can save it by clicking the **Disk** icon in the **Tour Control** tool. This allows you to delete the path and just retain the tour for subsequent use.

Excellent! You have just simulated the pyroclastic flow that destroyed Vesuvius. It is interesting to note that the way the tour "erupts" from the volcano, how it bounces along the terrain toward Pompeii, and the speed at which the tour progresses is a reasonable facsimile of an actual pyroclastic flow.

Project 4.3   creating tours from the Google Earth viewer

Not only can you save tours of placemarks and paths, but you also can record tours of the Google Earth viewer.

1. To create a tour, click on the **Record Tour** icon . The **Record Tour** control appears in the lower left corner of the Google Earth viewer window (Figure 4.6).

Figure 4.6: Record Tour control.

2. To turn tour recording on, click the red **Record** button. Anything that occurs in the Google Earth viewer will be recorded. To record audio during the tour, click the blue **Record Audio** button (microphone). When you are finished recording, click the red **Record** button again. The **Tour Control** (Figure 4.7) appears, and your recorded tour will play. You can save the tour to your **My Places** folder in the **Places** panel by clicking the **Disk** icon in the **Tour Control**.

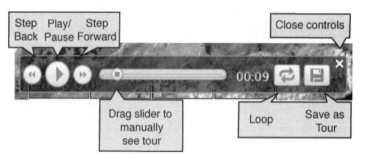

Figure 4.7: Tour Control.

That's it! You can now make recordings, save them to your **My Places** folder, and record audio narration to accompany the tour.

Project 4.4　creating animations of historical imagery I

1. In the **Fly to** text field in the **Search** panel, type 18.543157, -72.338851, and click on the **Magnifying Glass** icon to fly to the National Palace in Port-au-Prince, Haiti (use an eye altitude of approximately 300 m). This area was devastated by a 7.0 earthquake on January 12, 2010.

2. Click on the **Historical Imagery** icon ![icon] to see pre-earthquake imagery. A **time span control** (Figure 4.8) will appear in the upper left corner of the Google Earth viewer.

Figure 4.8: Time span control for historical imagery.

3. On the time span control, click the **button containing the left-pointing triangle (View Older Image** button) until the imagery date reads "8/26/2009". This is a pre-earthquake image of the area around the National Palace (Figure 4.9). *Please note that imagery dates in Google Earth are displayed according to user preference, so your dates may not exactly match the dates displayed in this workbook. Because imagery dates are stored in UTC (Coordinated Universal Time), the dates shown are more consistent if you click the* **Wrench** *icon (Figure 4.8) and set the date display to* **UTC***.*

*Figure 4.9: August 26, 2009, pre-earthquake image of the
National Palace in Port-au-Prince, Haiti.*

4. Now click the **button containing the right-pointing triangle (Step Forward** button) or drag the slider until the Imagery Date reads "1/17/2010". This shows a post-earthquake image of the area around the National Palace (Figure 4.10) a few days after the earthquake happened.

138

Figure 4.10: January 17, 2010, post-earthquake image of the National Palace in Port-au-Prince, Haiti.

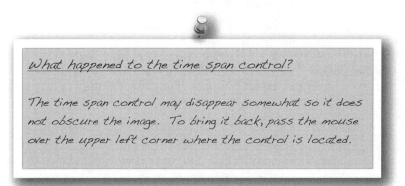

What happened to the time span control?

The time span control may disappear somewhat so it does not obscure the image. To bring it back, pass the mouse over the upper left corner where the control is located.

That is all there is to accessing Google Earth's historical imagery. The time range and image quality vary substantially from location to location. In the instance of the Haiti earthquake tragedy, the January 17, 2010 image is extremely high resolution and was beneficial in helping relief efforts in the region. In the next project, we'll investigate another example where historical imagery tells an important story.

Project 4.5   creating animations of historical imagery II

1. In the **Fly to** text field in the **Search** panel, type -65.37, -60.95, and click on the **Magnifying Glass** icon to fly to the Larsen B ice shelf in Antarctica. Make sure that your eye altitude is about 104 km.

2. Click on the **Historical Imagery** icon 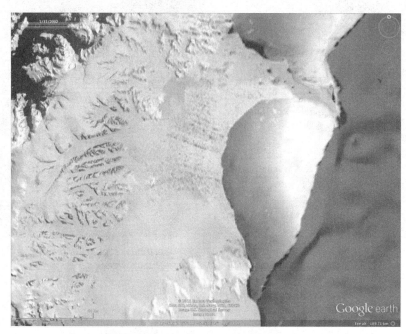 and set the time span control (Figure 4.8) to "1/31/2002" *(please note that Google currently has some display issues on some dates and at higher elevations)*. Figure 4.11 shows the Larsen B ice shelf with some ponding of water on the surface.

Figure 4.11: January 31, 2002 image of the Larsen B ice shelf in Antarctica.

3. Now click the **button containing the right-pointing triangle (Step Forward** button) or drag the slider until the Imagery Date reads "3/7/2002" (Figure 4.12) and then "2/23/2006" (Figure 4.13). This sequence of images from 2002 until 2006 captures the breakup of the Larsen B ice shelf.

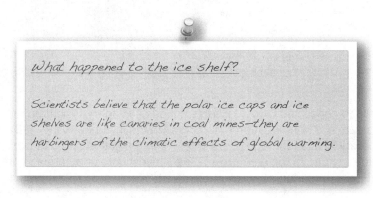

What happened to the ice shelf?

Scientists believe that the polar ice caps and ice shelves are like canaries in coal mines—they are harbingers of the climatic effects of global warming.

*Figure 4.12: March 7, 2002 image of the
Larsen B ice shelf in Antarctica.*

*Figure 4.13: February 23, 2006 image of the
Larsen B ice shelf in Antarctica.*

Leftovers

Leftovers are those last remaining tidbits of "good stuff" that are just too good to throw out (so we grouped them together in this module)! Some of our leftovers are a bit more advanced and require editing KML files using a text editor, but hopefully they will help you make the transition into additional programming/exploration with KML files on your own.

Project 5.1   importing GPS data into Google Earth

1. GPS data can be imported into Google Earth in three basic ways:

   - **Importing a .gpx data file directly into Google Earth**.

   - **Importing data directly from a GPS device** (works for most Garmin and Magellan GPS units).

   - **Using a third-party application to create a KML file from a .gpx data file or from imported data directly from your GPS device** (e.g., http://www.gpsbabel.org/ or http://www.gpsvisualizer.com/. *Please note that instructions for using third-party applications are beyond the scope of this workbook (both sites provide excellent tutorials).*

2. Connecting your GPS unit to a computer.

   a. If you plan on downloading GPS data on a Windows computer, first install the USB driver that comes with your GPS unit. Garmin GPS users should be able to download this driver from the Garmin website: http://www8.garmin.com/support/download.jsp. For Magellan GPS users, the download support site is http://support.magellangps.com/support/index.php?_m=downloads&_a=view.

   b. With the GPS *off*, hook the GPS unit to the computer running Google Earth using the provided cable. Turn the GPS *on*.

3. Within Google Earth, select **Tools > GPS**, and the **GPS Import** dialog appears (Figure 5.1).

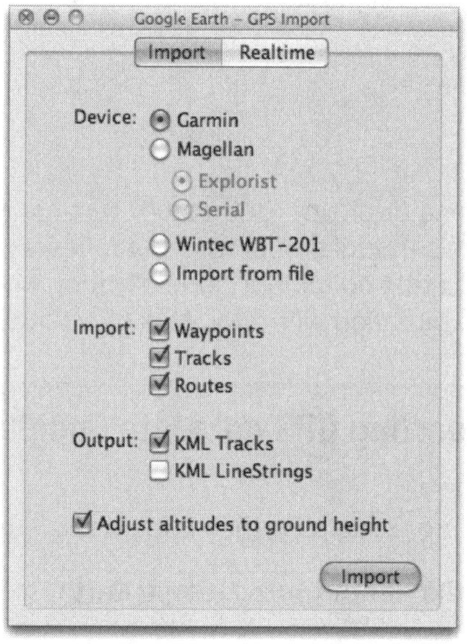

Figure 5.1: GPS Import dialog for downloading data directly from the GPS unit.

4. Select the appropriate **Device** (**Garmin**, **Magellan**, or **Import from File**).

5. Select the data types to **Import**:

   a. **Waypoints**—specific point locations recorded by the GPS user. Waypoints are represented in Google Earth as placemarks.

   b. **Tracks**—series of points automatically recorded by the GPS unit at a specified interval (sometimes referred to as breadcrumb trails). Tracks are represented in Google Earth as paths.

   c. **Routes**—series of points used by the GPS in routing (e.g., like using the "go-to" instructions). Routes are represented in Google Earth as paths.

6. In the **Output** section of the **GPS Import** dialog (Figures 5.1 and 5.2), users can choose:

   a. **KML Tracks**—Paths for imported **Tracks** and **Routes** will include a time element that can be animated as described in Projects 4.4 and 4.5.

   b. **KML LineStrings**—Paths for imported **Tracks** and **Routes** will not include a time element. Points comprising the path will be displayed and will have a time element.

7. A checkbox allows you to **Adjust altitudes to ground height**, which automatically converts all imported data to lie on the ground surface in the Google Earth viewer.

8. When you click **Import**, a dialog appears detailing the data imported into Google Earth (if you are doing **Import from File**, you will first see a system dialog box that allows you to browse for the file). The data will be stored in your **Temporary Places** folder in the **Places** panel within a folder "garmin GPS device" or similar (Figure 5.2). These data must be moved to **My Places**, or they will be lost when you quit Google Earth.

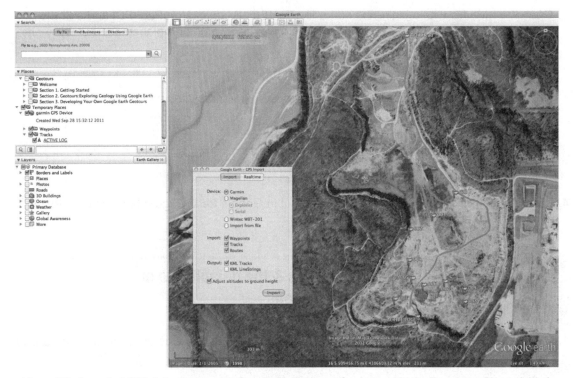

Figure 5.2: Imported GPS data displayed in Google Earth viewer (and stored in the Temporary Places folder).

Project 5.2   hiding description text in the Places panel

1. Click RMB on the feature (either in the **Places** panel or in the Google Earth viewer). Select **Save Place As...**, and a standard **Save** dialog appears.

2. Select **KML** from the **Format** drop-down menu.

3. Open the KML file with a plain-text editor like TextEdit (Mac) or NotePad (PC) and edit the feature's KML code as follows (this example shows the syntax for a placemark, but will work for other features as well):

```
<Placemark>
    <snippet maxLines="0"></snippet>  (the maxLines part is optional)
    ...
</Placemark>
```

4. Save the file with a .kml extension and reopen it in Google Earth (Figure 5.3; see bold text in example KML code provided after Project 5.4).

Project 5.3   hiding directions & the default title in placemark balloons

1. Click RMB on the placemark (either in the **Places** panel or in the Google Earth viewer). Select **Save Place As...**, and a standard **Save** dialog appears.

2. Select **KML** from the **Format** drop-down menu.

3. Open the KML file with a plain-text editor like TextEdit (Mac) or NotePad (PC) and edit the feature's KML code as follows: (*Note: This code must be inside all **Style** tags in the file.*)

```
<Style>
    ...
    </IconStyle>
    <BalloonStyle>
        <text>$[description]</text>
    </BalloonStyle>
</Style>
```

4. Save the file with a .kml extension and reopen it in Google Earth (Figure 5.3; see bold text in example KML code provided after Project 5.4).

Project 5.4 — inserting comments in KML files or descriptions

1. Click RMB on the feature (either in the **Places** panel or in the Google Earth viewer). Select **Save Place As...**, and a standard **Save** dialog appears.

2. Select **KML** from the **Format** drop-down menu.

3. Open the KML file with a plain-text editor like TextEdit (Mac) or NotePad (PC) and edit the feature's KML code as follows:

   <!-- insert comment here -->

> *Can comments be added to descriptions?*
>
> *Absolutely! Just click RMB on the feature in the Places panel and select Get Info (Mac) or Properties (PC). In the Description tab, you can add comments with the same syntax as described in Project 5.4. See Figure 1.7 for an example.*

4. Save the file with a .kml extension and reopen it in Google Earth (see bold text in example KML code provided after this project).

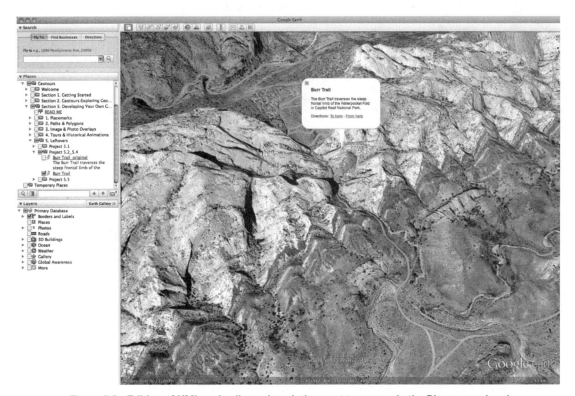

Figure 5.3: Editing of KML code allows descriptions not to appear in the Places panel and omits default titles and directions from placemark balloons.

Projects 5.2–5.4 example code

```xml
<?xml version="1.0" encoding="UTF-8"?>
<kml xmlns="http://www.opengis.net/kml/2.2" xmlns:gx="http://www.google.com/kml/ext/2.2" xmlns:kml="http://www.opengis.net/kml/
2.2" xmlns:atom="http://www.w3.org/2005/Atom">
<Document>
        <name>Description.kml</name>
        <StyleMap id="msn_ylw-pushpin">
                <Pair>
                        <key>normal</key>
                        <styleUrl>#sn_ylw-pushpin</styleUrl>
                </Pair>
                <Pair>
                        <key>highlight</key>
                        <styleUrl>#sh_ylw-pushpin</styleUrl>
                </Pair>
        </StyleMap>
        <Style id="sh_ylw-pushpin">
                <IconStyle>
                        <scale>1.3</scale>
                        <Icon>
                                <href>http://maps.google.com/mapfiles/kml/pushpin/ylw-pushpin.png</href>
                        </Icon>
                        <hotSpot x="20" y="2" xunits="pixels" yunits="pixels"/>
                </IconStyle>
        </Style>
        <Style id="sn_ylw-pushpin">
                <IconStyle>
                        <scale>1.1</scale>
                        <Icon>
                                <href>http://maps.google.com/mapfiles/kml/pushpin/ylw-pushpin.png</href>
                        </Icon>
                        <hotSpot x="20" y="2" xunits="pixels" yunits="pixels"/>
                </IconStyle>
                <BalloonStyle>
                        <text>$[description]</text> <!-- Project 5.3-hides directions & the default title in placemark balloons -->
                </BalloonStyle>
        </Style>
        <Placemark>  <!-- Project 5.4-comments for Projects 5.2 & 5.3 -->
                <snippet maxLines="0"></snippet>  <!-- Project 5.2-hides description text in the Places panel -->
                <name>Project 5.2_5.4_Burr Trail</name>
                <description>The Burr Trail traverses the steep frontal limb of the Waterpocket Fold in Capitol Reef National Park.</
description>
                <LookAt>
                        <longitude>-111.0184541629349</longitude>
                        <latitude>37.85141955131326</latitude>
                        <altitude>0</altitude>
                        <heading>-94.71699129715726</heading>
                        <tilt>41.05223703428219</tilt>
                        <range>1653.602789638165</range>
                        <altitudeMode>relativeToGround</altitudeMode>
                        <gx:altitudeMode>relativeToSeaFloor</gx:altitudeMode>
                </LookAt>
                <styleUrl>#msn_ylw-pushpin</styleUrl>
                <Point>
                        <altitudeMode>clampToGround</altitudeMode>
                        <gx:altitudeMode>clampToSeaFloor</gx:altitudeMode>
                        <coordinates>-111.0226329570657,37.85112675782791,0</coordinates>
                </Point>
        </Placemark>
</Document>
</kml>
```

Project 5.5   programming time animations

1. In the **Places** panel, click RMB on the folder containing the features that you want to animate. Select **Save Place As...**, and a standard **Save** dialog appears.

2. Select **KML** from the **Format** drop-down menu.

3. Open the KML file with a plain-text editor like TextEdit (Mac) or NotePad (PC). Add the **TimeSpan** tags with **begin** and **end** tags after the **description** tag of each feature. If a feature is to be left on until present (i.e., activated at a certain time and then left on), it does not need an end tag.

4. Save the file with a .kml extension and reopen it in Google Earth. When the folder is checked, the time span control will appear in the upper lefthand corner so that you can view folder items based on their time stamps.

```
<Folder>
   ...
   </description>
   <TimeSpan>
      <begin>-Year-Month-Day</begin>
      <end>-Year-Month-Day</end>
   </TimeSpan>
   ...
</Folder>
```

5. In the KML example below, three placemarks were created and placed in a folder. The three placemarks highlight rock units in the Henry Mountains and Capitol Reef National Park areas that were formed in the Cenozoic, Mesozoic, and Paleozoic Eras, respectively. All of the KML code was automatically generated by Google Earth, except for the addition of the three TimeSpan entries (bold). For this example, we used the years to represent millions of years and a negative sign (-) to denote the past. Each placemark will appear and disappear as specified by its begin and end tags. A placemark first will appear during the period from 542 to 245 million years ago (~Paleozoic). Then the Paleozoic placemark will disappear, and a second placemark will appear during the period from 245 to 65 million years ago (~Mesozoic). Lastly, the Mesozoic placemark will disappear, and a third placemark will appear during the period from 65 to 1 million years ago (~Cenozoic).

149

Project 5.5   example code

```
<?xml version="1.0" encoding="UTF-8"?>
<kml xmlns="http://www.opengis.net/kml/2.2" xmlns:gx="http://www.google.com/kml/ext/2.2" xmlns:kml="http://www.opengis.net/kml/2.2" xmlns:atom="http://
www.w3.org/2005/Atom">
<Document>
        <name>Project5.kml</name>
        <StyleMap id="msn_ylw-pushpin">
                <Pair>
                        <key>normal</key>
                        <styleUrl>#sn_ylw-pushpin0</styleUrl>
                </Pair>
                <Pair>
                        <key>highlight</key>
                        <styleUrl>#sh_ylw-pushpin</styleUrl>
                </Pair>
        </StyleMap>
        <Style id="sn_ylw-pushpin">
                <IconStyle>
                        <scale>1.1</scale>
                        <Icon>
                                <href>http://maps.google.com/mapfiles/kml/pushpin/ylw-pushpin.png</href>
                        </Icon>
                        <hotSpot x="20" y="2" xunits="pixels" yunits="pixels"/>
                </IconStyle>
        </Style>
        <Style id="sh_ylw-pushpin">
                <IconStyle>
                        <scale>1.3</scale>
                        <Icon>
                                <href>http://maps.google.com/mapfiles/kml/pushpin/ylw-pushpin.png</href>
                        </Icon>
                        <hotSpot x="20" y="2" xunits="pixels" yunits="pixels"/>
                </IconStyle>
        </Style>
        <Style id="sh_ylw-pushpin0">
                <IconStyle>
                        <scale>1.3</scale>
                        <Icon>
                                <href>http://maps.google.com/mapfiles/kml/pushpin/ylw-pushpin.png</href>
                        </Icon>
                        <hotSpot x="20" y="2" xunits="pixels" yunits="pixels"/>
                </IconStyle>
        </Style>
        <Style id="sh_ylw-pushpin1">
                <IconStyle>
                        <scale>1.3</scale>
                        <Icon>
                                <href>http://maps.google.com/mapfiles/kml/pushpin/ylw-pushpin.png</href>
                        </Icon>
                        <hotSpot x="20" y="2" xunits="pixels" yunits="pixels"/>
                </IconStyle>
        </Style>
        <StyleMap id="msn_ylw-pushpin0">
                <Pair>
                        <key>normal</key>
                        <styleUrl>#sn_ylw-pushpin</styleUrl>
                </Pair>
                <Pair>
                        <key>highlight</key>
                        <styleUrl>#sh_ylw-pushpin0</styleUrl>
                </Pair>
        </StyleMap>
        <Style id="sn_ylw-pushpin0">
                <IconStyle>
                        <scale>1.1</scale>
                        <Icon>
                                <href>http://maps.google.com/mapfiles/kml/pushpin/ylw-pushpin.png</href>
                        </Icon>
                        <hotSpot x="20" y="2" xunits="pixels" yunits="pixels"/>
                </IconStyle>
        </Style>
        <StyleMap id="msn_ylw-pushpin1">
                <Pair>
                        <key>normal</key>
                        <styleUrl>#sn_ylw-pushpin1</styleUrl>
                </Pair>
                <Pair>
                        <key>highlight</key>
                        <styleUrl>#sh_ylw-pushpin1</styleUrl>
                </Pair>
        </StyleMap>
```

150

```
<Style id="sn_ylw-pushpin1">
        <IconStyle>
                <scale>1.1</scale>
                <Icon>
                        <href>http://maps.google.com/mapfiles/kml/pushpin/ylw-pushpin.png</href>
                </Icon>
                <hotSpot x="20" y="2" xunits="pixels" yunits="pixels"/>
        </IconStyle>
</Style>
<Folder>
        <name>Project 5.5</name>
        <visibility>0</visibility>
        <Placemark>
                <name>Cenozoic</name>
                <visibility>0</visibility>
                <description>Tertiary intrusion that formed the Henry Mountains in Utah.</description>
                <TimeSpan>
                        <begin>-065-12-31</begin>
                        <end>-001-12-31</end> <!--don't need an end tag if it goes to present-->
                </TimeSpan>
                <LookAt>
                        <longitude>-110.9554208566362</longitude>
                        <latitude>37.9176967512585</latitude>
                        <altitude>0</altitude>
                        <heading>-27.44322278882175</heading>
                        <tilt>45.82052918099446</tilt>
                        <range>42374.69977423045</range>
                        <altitudeMode>relativeToGround</altitudeMode>
                        <gx:altitudeMode>relativeToSeaFloor</gx:altitudeMode>
                </LookAt>
                <styleUrl>#msn_ylw-pushpin</styleUrl>
                <Point>
                        <altitudeMode>clampToGround</altitudeMode>
                        <gx:altitudeMode>clampToSeaFloor</gx:altitudeMode>
                        <coordinates>-110.7988909377261,37.94668383465456,0</coordinates>
                </Point>
        </Placemark>
        <Placemark>
                <name>Mesozoic</name>
                <visibility>0</visibility>
                <description>Jurassic Navajo Sandstone along the steep frontal limb of the Waterpocket Fold.</description>
                <TimeSpan>
                        <begin>-245-12-31</begin>
                        <end>-065-12-31</end>
                </TimeSpan>
                <LookAt>
                        <longitude>-110.9554208566362</longitude>
                        <latitude>37.9176967512585</latitude>
                        <altitude>0</altitude>
                        <heading>-27.44322278882175</heading>
                        <tilt>45.82052918099446</tilt>
                        <range>42374.69977423045</range>
                        <altitudeMode>relativeToGround</altitudeMode>
                        <gx:altitudeMode>relativeToSeaFloor</gx:altitudeMode>
                </LookAt>
                <styleUrl>#msn_ylw-pushpin0</styleUrl>
                <Point>
                        <altitudeMode>clampToGround</altitudeMode>
                        <gx:altitudeMode>clampToSeaFloor</gx:altitudeMode>
                        <coordinates>-111.0237276810503,37.85352295017488,0</coordinates>
                </Point>
        </Placemark>
        <Placemark>
                <name>Paleozoic</name>
                <visibility>0</visibility>
                <description>Stream valleys cut down to expose the Paleozoic Kaibab Limestone. </description>
                <TimeSpan>
                        <begin>-542-12-31</begin>
                        <end>-245-12-31</end>
                </TimeSpan>
                <LookAt>
                        <longitude>-110.9554208566362</longitude>
                        <latitude>37.9176967512585</latitude>
                        <altitude>0</altitude>
                        <heading>-27.44322278882175</heading>
                        <tilt>45.82052918099446</tilt>
                        <range>42374.69977423045</range>
                        <altitudeMode>relativeToGround</altitudeMode>
                        <gx:altitudeMode>relativeToSeaFloor</gx:altitudeMode>
                </LookAt>
                <styleUrl>#msn_ylw-pushpin1</styleUrl>
                <Point>
                        <altitudeMode>clampToGround</altitudeMode>
                        <gx:altitudeMode>clampToSeaFloor</gx:altitudeMode>
                        <coordinates>-111.0471873204594,37.82153322522905,0</coordinates>
                </Point>
        </Placemark>
</Folder>
</Document>
</kml>
```

151

Project 5.6   finding useful Google Earth projects on the Web

- Click the **Earth Gallery** button in the **Layers** panel. Click the **Back to Google Earth** button in the upper left corner of the Google Earth viewer to return to Google Earth.

- Digital Geography—http://www.digitalgeography.co.uk/

- Earthswoop—http://www.earthswoop.com/

- Google Earth Blog—http://www.gearthblog.com/

- Google Earth Community—http://bbs.keyhole.com/ubb/ubbthreads.php/Cat/0

- Google Earth Cool Places—http://googleearthcoolplaces.com/

- Google Earth Design—http://googleearthdesign.blogspot.com/

- Google Earth Guidebook—http://google-earth.guide-book.co.uk/

- Google Earth Hacks—http://www.gearthhacks.com/

- Google Earth Lessons—http://www.gelessons.com/

- Google Lat Long Blog—http://google-latlong.blogspot.com/

- Google Maps Mania—http://googlemapsmania.blogspot.com/

- Google Sightseeing—http://googlesightseeing.com/

- Ogle Earth—http://ogleearth.com/

- Teaching with Google Earth—http://serc.carleton.edu/sp/library/google_earth/index.html

- USGS Earthquake data—http://earthquake.usgs.gov/learn/kml.php

Section 1: Getting Started
Part 1
Part 2

Section 2: Exploring Geology Using Google Earth
A. Earth & Sky
- Chesapeake Bay Impact overlay-Marshak, S., 2008, Earth: Portrait of a Planet, 3rd ed., W.W. Norton, 832 p.
- Impact Model image-Marshak, S., 2009, Essentials of Geology, 3rd ed., W.W. Norton, 518 p.
B. Plate Tectonics
- Iceland Geology overlay-Marshak, S., 2008, Earth: Portrait of a Planet, 3rd ed., W.W. Norton, 832 p.
- Seafloor Age Map overlay-Seafloor age data downloaded from http://www.earthbyte.org/resources.html and made into a Google Earth overlay by Dr. Tim Cope, DePauw University. Original citation for seafloor age data: Müller, R.D., Roest, W.R., Royer, J.-Y., Gahagan, L.M., and Sclater, J.G., A digital age map of the ocean floor. SIO Reference Series 93-30, Scripps Institution of Oceanography, http://www.ngdc.noaa.gov/mgg/global/& crustage.HTML.
- Earthquake overlay-Earthquake data downloaded fromhttp://earthquake.usgs.gov/regional/neic/and color-coded by depth (km) to create a Google Earth overlay by Dr. Tim Cope, DePauw University.
- Hawaiian Island ages-Problem adapted from a similar exercise developed by Dr. Tim Cope, DePauw University. Data adapted from a web site by Ken Rubin (2005). Additional information available at:http://www.soest.hawaii.edu/ GG/HCV/haw_formation.html. Additional information at:http://earthquake.usgs.gov/regional/neic/
- Sag Pond image-Marshak, S., 2009, Essentials of Geology, 3rd ed., W.W. Norton, 518 p.
- Offset Stream image-Marshak, S., 2009, Essentials of Geology, 3rd ed., W.W. Norton, 518 p.
- Hot Spot Trail Map overlay-Marshak, S., 2008, Earth: Portrait of a Planet, 3rd ed., W.W. Norton, 832 p.
- Yellowstone Calderas overlay-Marshak, S., 2008, Earth: Portrait of a Planet, 3rd ed., W.W. Norton, 832 p.
C. Minerals
D. Igneous Rocks
- North America Batholiths Map overlay-Marshak, S., 2008, Earth: Portrait of a Planet, 3rd ed., W.W. Norton, 832 p.
E. Volcanoes
- Mt. St. Helens Volcanic Features overlay-Marshak, S., 2009, Essentials of Geology, 3rd ed., W.W. Norton, 518 p.
- 79 C.E. Eruption Time Sequence overlay-Adapted from Foss, P, Sigurdsson, H., and Robertson, S., 2007, Map 1 and Fig. 4.4, in J.J. Dobbins and P.W. Foss (eds.), The World of Pompeii, Routledge, 704 pages. More information at: http://homepage.mac.com/pfoss/Pompeii/WorldOfPompeii/index.html and http:// www.routledgearchaeology.com/books/The-World-of-Pompeii-isbn9780415475778.
- Hawaiian Island Landslides overlay-Marshak, S., 2008, Earth: Portrait of a Planet, 3rd ed., W.W. Norton, 832 p. (adapted from USGS/Barry W. Eakins).
- Mt. Rainier Hazards Map overlay-Marshak, S., 2008, Earth: Portrait of a Planet, 3rd ed., W.W. Norton, 832 p. (adapted from Fact Sheet by Scott, K.M., Wolfe, E.W., and Driedger, C.L., Hawaiian Volcano Observatory/U.S. Geological Survey).
- Mt. Rainier Features Map overlay-Overlay courtesy of the USGS (Topinka, USGSICVO, 1997; modified from Scott, et al., 1992). More information at: http://vulcan.wr.usgs.gov/Volcanoes/Rainier/Maps/map_place_names.html.
- Crater Lake, OR Map overlay-Overlay courtesy of the USGS (Ramsey, D.W., Dartnell, P., Bacon, C.R., Robinson, J.E., and Gardner, 2003, Crater Lake Revealed, U.S. Geological Survey Geologic Investigations Series I-2790). More information at: http://geopubs.wr.usgs.gov/i-map/i2790/.
- Contour Map overlays-Adapted from Google Maps.
- Pyroclastic Flow video-Video courtesy of Miyagi, I., Kawanabe, Y., Takada, A., Sakaguchi, K., Takarada, S., 2007, A Collection of Unzen Video Clips, GSJ Open-file Report no. 469, Geological Survey of Japan, AIST. More information at: http://www.gsj.jp/GDB/openfile/files/no0469/contents0469/index.html.
F. Sedimentary Rocks
- Grand Canyon Geologic Map overlay-Overlay courtesy of the USGS (Billingsley, G.H., 2000, Geologic Map of the Grand Canyon 30' x 60' Quadrangle, Coconino and Mohave Counties, Northwestern Arizona, U.S. Geological Survey, Geologic Investigations Series I-2688, Version 1.0.). More information at: http://pubs.usgs.gov/imap/ i-2688/.
- Great Unconformity video-Video courtesy of: Wilkerson (2011).

G. Metamorphic Rocks

- Wind River Mts., WY Geologic Map overlay-Geologic map overlay generated using Google Earth application downloaded from http://www.gelib.com/geologic-map-united-states.htm. Original citation for entire geologic map: King, P.B., Beikman, H.M., and Edmonston, G.J. 1974, Geology of the Conterminous United States at 1:2,500,000 Scale, United States Geological Survey, http://pubs.usgs.gov/dds/dds11/index.html.
- New England Metamorphic Zones overlay-Marshak, S., 2008, Earth: Portrait of a Planet, 3rd ed., W.W. Norton, 832 p.
- Global Map of Shields & Mt Belts overlay-Marshak, S., 2008, Earth: Portrait of a Planet, 3rd ed., W.W. Norton, 832 p.

H. Earthquakes

- Regional Plate Tectonics overlay-Marshak, S., 2009, Essentials of Geology, 3rd ed., W.W. Norton, 518 p.
- New Madrid, MO Seismicity overlay-Marshak, S., 2008, Earth: Portrait of a Planet, 3rd ed., W.W. Norton, 832 p.
- Sag Pond image-Marshak, S., 2009, Essentials of Geology, 3rd ed., W.W. Norton, 518 p.
- Offset Stream image-Marshak, S., 2009, Essentials of Geology, 3rd ed., W.W. Norton, 518 p.
- 2004 Tsunami Travel-Time Animation image-Image courtesy of Kenji Satake at the Active Fault Research Center in Tsukuba, Japan, http://staff.aist.go.jp/kenji.satake/animation.gif.
- 2004 Epicenter image-Marshak, S., 2009, Essentials of Geology, 3rd ed., W.W. Norton, 518 p.
- North Anatolian Fault image-Marshak, S., 2009, Essentials of Geology, 3rd ed., W.W. Norton, 518 p. (adapted from Stein et al., 1996).

I. Geologic Structures

- Geologic Map of the Bow Corridor, Canada overlay-Hamilton, W.N., Price, M.C., and Chao,D.K.,1998, Geology and mineral deposits of the Bow Corridor. Map 232, Alberta Geological Survey. http://www.ags.gov.ab.ca/publications/ ABSTRACTS/MAP_232.html. Used with permission from ERCB/AGS.
- Calgary to Castle Mtn Geologic Map, Canada overlay-Ollerenshaw, N.C.,1978, Geology, Calgary, West of Fifth Meridian, Alberta-British Columbia. Map 1457A, Geological Survey of Canada. http://apps1.gdr.nrcan.gc.ca/mirage/ show_image_e.php?client=mrsid2&id=1091003=gscmap-a_1457a_e_1978_mn01.sid. Reproduced with the permission of Natural Resources Canada 2009, courtesy of the Geological Survey of Canada (Map 1457A).
- Kananaskis Country Geologic Map, Canada overlay-McMechan, M.E.,1995, Rocky Mountain Foothills and Front Ranges in Kananaskis Country, West of Fifth Meridian, Alberta. Map 1865A, Geological Survey of Canada. http:// apps1.gdr.nrcan.gc.ca/mirage/show_image_e.php?client=mrsid2&id=2048973=gscmap- a_1865a_e_1995_mg01.sid. Reproduced with the permission of Natural Resources Canada 2009, courtesy of the Geological Survey of Canada (Map 1865A).
- Sheep Mountain Geologic Map overlay-Map reprinted by permission of the AAPG whose permission is required for further use (AAPG © 2004): Banerjee, S., and Mitra, S., 2004, Remote surface mapping using orthophotos and geologic maps draped over digital elevation models: Application to the Sheep Mountain anticline, Wyoming: AAPG Bulletin, v. 88, no. 9, 1227-1237. Formation contacts modified from Hennier (1984). Black strike & dip data are from Hennier (1984), whereas white strike & dip values are interpreted from DEM data by Banerjee & Mitra (2004).
- Split Mountain Anticline, UT Geologic Map overlay-Geologic map georeferenced in Google Earth by Dr. Tim Cope, DePauw University. Derived from Rowley, P.D. and Hansen, W.R., 1979, Geologic map of the Split Mountain quadrangle, Uintah County, Utah: U.S. Geological Survey, Geologic Quadrangle Map GQ-1515, scale 1:24000, and Rowley, P.D., Kinney, D.M., and Hansen, W.R., 1979, Geologic map of the Dinosaur Quarry quadrangle, Uintah County, Utah: U.S. Geological Survey, Geologic Quadrangle Map GQ-1513, scale 1:24000. Additional information at: http://pubs.er.usgs.gov/usgspubs/gq/gq1515.
- Geologic Map of Quail Creek SP, UT overlay-Original citation (with permission from the Utah Geological Survey): Biek, R. F., 2000, Geology of Quail Creek State Park, Utah, in Geology of Utah's Parks and Monuments, Sprinkel, D. A., Chidsey, T. C., and Anderson, P. B., eds., Utah Geological Association Publication 28. http:// www.utahgeology.org/uga_publist.htm.
- Arches NP Geologic Map-North overlay-Original citation (with permission from the Utah Geological Survey): Doelling, H. H., 2000, Geology of Arches National Park, Grand County, Utah, in Geology of Utah's Parks and Monuments, Sprinkel, D. A., Chidsey, T. C., and Anderson, P. B., eds., Utah Geological Association Publication 28. http://www.utahgeology.org/uga_publist.htm.
- Arches NP Geologic Map-South overlay-Original citation (with permission from the Utah Geological Survey): Doelling, H. H., 2000, Geology of Arches National Park, Grand County, Utah, in Geology of Utah's Parks and Monuments, Sprinkel, D. A., Chidsey, T. C., and Anderson, P. B., eds., Utah Geological Association Publication 28. http://www.utahgeology.org/uga_publist.htm.
- Geologic Map of Canyonlands NP overlay-Original citation (with permission from the Utah Geological Survey): Baars, D. L., 2000, Geology of Canyonlands National Park, Utah, in Geology of Utah's Parks and Monuments, Sprinkel, D. A., Chidsey, T. C., and Anderson, P. B., eds., Utah Geological Association Publication 28. http:// www.utahgeology.org/uga_publist.htm.
- Fin Garden video-Video courtesy of Wilkerson, M.S., 2011, personal videos.

J. Geologic Time

- Grand Canyon Geologic Map overlay-Overlay courtesy of the USGS (Billingsley, G.H., 2000, Geologic Map of the Grand Canyon 30' x 60' Quadrangle, Coconino and Mohave Counties, Northwestern Arizona, U.S. Geological Survey, Geologic Investigations Series I-2688, Version 1.0.). More information at: http://pubs.usgs.gov/imap/i-2688/.
- Geologic Map of Zion NP overlay-Original citation (with permission from the Utah Geological Survey): Biek, R. F., Willis, G. C., Hylland, M. D., and Doelling, H. H., 2000, Geology of Zion National Park, Utah, in Geology of Utah's Parks and Monuments, Sprinkel, D. A., Chidsey, T. C., and Anderson, P. B., eds., Utah Geological Association Publication 28. http://www.utahgeology.org/uga_publist.htm.
- Geologic Map of Cedar Breaks NM, UT overlay-Original citation (with permission from the Utah Geological Survey): Hatfield, S.C., and others, 2000, Geology of Cedar Breaks National Monument, Utah, in Geology of Utah's Parks and Monuments, Sprinkel, D. A., Chidsey, T. C., and Anderson, P. B., eds., Utah Geological Association Publication 28. http://www.utahgeology.org/uga_publist.htm.
- Geologic Map of Grand Staircase, UT overlay-Original citation (with permission from the Utah Geological Survey): Doelling, H. H.,, and others, 2000, Geology of Grand Staircase-Escalante National Monument, Utah, in Geology of Utah's Parks and Monuments, Sprinkel, D. A., Chidsey, T. C., and Anderson, P. B., eds., Utah Geological Association Publication 28. http://www.utahgeology.org/uga_publist.htm.
- Geologic Map of the Kaiparowits Plateau, UT overlay-Original citation (with permission from the Utah Geological Survey): Doelling, H. H.,, and others, 2000, Geology of Grand Staircase-Escalante National Monument, Utah, in Geology of Utah's Parks and Monuments, Sprinkel, D. A., Chidsey, T. C., and Anderson, P. B., eds., Utah Geological Association Publication 28. http://www.utahgeology.org/uga_publist.htm.
- Geologic Map of Circle Cliffs, UT overlay-Original citation (with permission from the Utah Geological Survey): Doelling, H. H.,, and others, 2000, Geology of Grand Staircase-Escalante National Monument, Utah, in Geology of Utah's Parks and Monuments, Sprinkel, D. A., Chidsey, T. C., and Anderson, P. B., eds., Utah Geological Association Publication 28. http://www.utahgeology.org/uga_publist.htm.
- Geologic Map of Quail Creek SP, UT overlay-Original citation (with permission from the Utah Geological Survey): Biek, R. F., 2000, Geology of Quail Creek State Park, Utah, in Geology of Utah's Parks and Monuments, Sprinkel, D. A., Chidsey, T. C., and Anderson, P. B., eds., Utah Geological Association Publication 28. http://www.utahgeology.org/uga_publist.htm.
- Geologic Map of Flaming Gorge, UT overlay-Original citation (with permission from the Utah Geological Survey): Sprinkel, D. A., 2000, Geology of Flaming Gorge National Recreation Area, UT-WY, in Geology of Utah's Parks and Monuments, Sprinkel, D. A., Chidsey, T. C., and Anderson, P. B., eds., Utah Geological Association Publication 28. http://www.utahgeology.org/uga_publist.htm.
- Geologic Map of Antelope Island SP, UT overlay-Original citation (with permission from the Utah Geological Survey): Willis, G. C., Yonkee, A., Doelling, H. H.,, and Jensen, M. E., 2000, Geology of Antelope Island State Park, Utah, in Geology of Utah's Parks and Monuments, Sprinkel, D. A., Chidsey, T. C., and Anderson, P. B., eds., Utah Geological Association Publication 28. http://www.utahgeology.org/uga_publist.htm.
- Geologic Map of Dead Horse Point SP, UT overlay-Original citation (with permission from the Utah Geological Survey): Doelling, H. H.,, Chidsey, Jr., T. C., and Benson, B. J., 2000, Geology of Dead Horse Point State Park, Grand and San Juan Counties, Utah, in Geology of Utah's Parks and Monuments, Sprinkel, D. A., Chidsey, T. C., and Anderson, P. B., eds., Utah Geological Association Publication 28. http://www.utahgeology.org/uga_publist.htm.
- Geologic Map of Canyonlands NP overlay-Original citation (with permission from the Utah Geological Survey): Baars, D. L., 2000, Geology of Canyonlands National Park, Utah, in Geology of Utah's Parks and Monuments, Sprinkel, D. A., Chidsey, T. C., and Anderson, P. B., eds., Utah Geological Association Publication 28. http://www.utahgeology.org/uga_publist.htm.
- Arches NP Geologic Map-North overlay-Original citation (with permission from the Utah Geological Survey): Doelling, H. H., 2000, Geology of Arches National Park, Grand County, Utah, in Geology of Utah's Parks and Monuments, Sprinkel, D. A., Chidsey, T. C., and Anderson, P. B., eds., Utah Geological Association Publication 28. http://www.utahgeology.org/uga_publist.htm.
- Arches NP Geologic Map-South overlay-Original citation (with permission from the Utah Geological Survey): Doelling, H. H., 2000, Geology of Arches National Park, Grand County, Utah, in Geology of Utah's Parks and Monuments, Sprinkel, D. A., Chidsey, T. C., and Anderson, P. B., eds., Utah Geological Association Publication 28. http://www.utahgeology.org/uga_publist.htm.
- Stratigraphy of the Grand Canyon image-Marshak, S., 2009, Essentials of Geology, 3rd ed., W.W. Norton, 518 p.
- Great Unconformity video-Video courtesy of Wilkerson, M.S., 2011, personal videos.
- Fin Garden video-Video courtesy of Wilkerson, M.S., 2011, personal videos.

K. Earth History

- Appalachian Mts., TN Geologic Map overlay-Geologic map overlay generated using Google Earth application downloaded from http://www.gelib.com/geologic-map-united-states.htm. Original citation for entire geologic map: King, P.B., Beikman, H.M., and Edmonston, G.J. 1974, Geology of the Conterminous United States at 1:2,500,000 Scale, United States Geological Survey, http://pubs.usgs.gov/dds/dds11/index.html.
- Global Paleogeographic Model overlay-Images created by Dr. Ron Blakey (images used with permission and cannot be used by others for commercial purposes without Dr. Blakey's permission). The time-animated Google Earth overlays were created by Tony Pack (Google Earth materials used with permission and cannot be used by others for commercial purposes without Tony Pack's permission).

L. Energy & Mineral Resources

- Major Known Oil Reserves polygons-Adapted from Marshak, S., 2009, Essentials of Geology, 3rd ed., W.W. Norton, 518 p.

M. Mass Movements

- Hawaiian Island Landslides overlay-Marshak, S., 2008, Earth: Portrait of a Planet, 3rd ed., W.W. Norton, 832 p. (adapted from USGS/Barry W. Eakins).
- Mt. St. Helens Volcanic Features overlay-Marshak, S., 2009, Essentials of Geology, 3rd ed., W.W. Norton, 518 p.

N. Stream Landscapes

- Channeled Scablands overlay-Marshak, S., 2008, Earth: Portrait of a Planet, 3rd ed., W.W. Norton, 832 p.
- Mississippi River Delta overlay-Marshak, S., 2008, Earth: Portrait of a Planet, 3rd ed., W.W. Norton, 832 p.

O. Oceans & Coastlines

- U.S. East Coast Sea Level Changes overlay-Marshak, S., 2008, Earth: Portrait of a Planet, 3rd ed., W.W. Norton, 832 p. (adapted from Kraft, 1973).

P. Groundwater & Karst Landscapes

- Everglades, FL-Previous Groundwater Flow overlay-Marshak, S., 2008, Earth: Portrait of a Planet, 3rd ed., W.W. Norton, 832 p.
- Everglades, FL-Present Groundwater Flow overlay-Marshak, S., 2008, Earth: Portrait of a Planet, 3rd ed., W.W. Norton, 832 p.
- Winter Park, FL sinkhole-Photo courtesy of the USGS. More information at:http://sofia.usgs.gov/publications/ofr/01-180/opandaction.html

Q. Desert Landscapes

- Syncline Development image-Marshak, S., 2008, Earth: Portrait of a Planet, 3rd ed., W.W. Norton, 832 p.

R. Glacial Landscapes

- Glacial Moraines-Long Island, NY & Cape Cod, MA overlay-Marshak, S., 2008, Earth: Portrait of a Planet, 3rd ed., W.W. Norton, 832 p. (adapted from Tarbuck & Lutgens, 1996).

S. Global Change

- Köppen-Geiger Climate Classification overlay-Overlay courtesy of Wilkerson, M.S., and Wilkerson, M.B., http://www.depauw.edu/acad/geosciences/mswilke/DELUGE.html. Original GIS shapefile downloaded from http://koeppen-geiger.vu-wien.ac.at/ from the Department of Natural Sciences, University of Veterinary Medicine Vienna. Original citation: Kottek, M., J. Grieser, C. Beck, B. Rudolf, and F. Rubel, 2006: World Map of the Kppen-Geiger climate classification updated. Meteorol. Z., 15, 259-263.DOI: 10.1127/0941-2948/2006/0130.

Section 3: Developing Your Own Interactive Google Earth Materials

1. Placemarks

- Mount Saint Helens image-Courtesy of Topinka, L., USGS/Cascades Volcano Observatory, http://vulcan.wr.usgs.gov/Imgs/Jpg/MSH/Images/MSH82_st_helens_plume_from_harrys_ridge_05-19-82_med.jpg.
- Mount Saint Helens video-Courtesy of USGS YouTube channel (http://www.youtube.com/user/usgs), United States Geological Survey, http://youtu.be/sC9JnuDuBsU.

2. Paths & Polygons

3. Image & Photo Overlays

- Chicxulub Crater Satellite Image overlay-Overlay courtesy of Short, Sr., N.M., NASA, http://rst.gsfc.nasa.gov/Sect18/Sect18_4.html.
- Chicxulub Crater Gravity Map overlay-Overlay courtesy of Short, Sr., N.M., NASA, http://rst.gsfc.nasa.gov/Sect18/Sect18_4.html.
- Half Dome photo overlay-Courtesy of Wilkerson, M.S., 2011, personal photos.

4. Tours & Historical Animations

- Tour of Folder Placemarks-Placemarks courtesy of Wilkerson, M.S., (2011). Derived from Powell, J.W., 1997, The Exploration of the Colorado River and Its Canyons, Penguin Books, 397 p.

5. Leftovers

E. Volcanoes

Marshak-See For Yourself Sites
Yellowstone Falls, WY
Mt. Saint Helens, WA-Vertical
Mt. Saint Helens, WA-Oblique
Smoking Volcano, Ecuador
Mt. Etna, Sicily-Vertical
Mt. Etna, Sicily-Oblique
Mt. Vesuvius, Italy
Hawaiian Volcanoes-Mauna Loa
Hawaiian Volcanoes-Kilauea
Mt. Shishildan, AK
East African Rift-Mt. Kilimanjaro
East African Rift-Goma Region
East African Rift-Goma Region Airport
Shield Volcanoes
Kilauea, HI
Mauna Loa, HI
Mauna Kea, HI
Diamond Head, HI
Composite Cone Volcanoes
Mt. Shishildan, AK
Mt. Rainier, WA
Mt. Saint Helens, WA
Crater Lake, OR
Mt. Shasta, CA
Tungurahua, Ecuador
Izalco Volcano, El Salvador-Vertical
Izalco Volcano, El Salvador-Oblique
Mt. Etna, Sicily-Vertical
Mt. Etna, Sicily-Oblique
Mt. Vesuvius, Italy
Cinder Cone Volcanoes
Sunset Crater, AZ
SP Mountain, AZ
Menan Buttes, ID
East African Rift-Mt. Kilimanjaro
East African Rift-Goma Region
East African Rift-Goma Region Airport
Obsidian Flow, Glass Mt., CA
Olympus Mons, Mars

F. Sedimentary Rocks

Marshak-See For Yourself Sites
Grand Canyon, AZ-Vertical
Grand Canyon, AZ-Oblique
Lewis Range, MT
Death Valley, CA
Great Exhuma, Bahamas
Sand Dunes, Namibia-Overview
Sand Dunes, Namibia-Detailed
Niger Delta, Nigeria-Overview
Niger Delta, Nigeria-Detailed
Grand Canyon, AZ
Niger Delta, Nigeria
Lewis Range, MT
Death Valley, CA
Great Exhuma, Bahamas
Sand Dunes, Namibia-Overview
Sand Dunes, Namibia-Detailed

G. Metamorphic Rocks

Marshak-See For Yourself Sites
Wind River Mts., WY-Overview
Wind River Mts., WY-Detailed
Canadian Shield, East of Hudson Bay
Pilbara Craton, Western Australia
Wind River Mts, WY
Canadian Shield, East of Hudson Bay
Pilbara Craton, Western Australia
New England Metamorphic Zones
Global Map of Shields & Mt Belts
Baraboo Quartzite, WI

H. Earthquakes

Marshak-See For Yourself Sites
San Andreas Fault-Overview
San Andreas Fault-Offset Stream
San Andreas Fault-Oblique
San Andreas Fault, San Francisco-Vertical
San Andreas Fault, San Francisco-Oblique
San Andreas Fault, San Francisco-Scarps
Salt Lake City, UT-Overview
Salt Lake City, UT-Wasatch Front-North
Salt Lake City, UT-Wasatch Front-South
Tsunami Damage, Banda Aceh-Overview
Tsunami Damage, Banda Aceh-Detailed (N)
Tsunami Damage, Banda Aceh-Detailed (S)
San Andreas Fault, CA
Salt Lake City, UT
New Madrid, MO Seismicity
2004 Sumatra Earthquake, Banda Aceh, Sumatra
North Anatolian Fault, Turkey

I. Geologic Structures

Marshak-See For Yourself Sites
Erosion in the Alps-Overview
Erosion in the Alps-Oblique
Mt. Everest-Overview
Mt. Everest-Oblique
Joints in Arches NP-Vertical
Joints in Arches NP-Oblique
Horsts & Grabens, Canyonlands
Alpine Fault, New Zealand
Faults & Folds, Makran Range
Appalachian Fold Belt, PA
Dome & Syncline, Western Australia
Faults
Normal
Horsts & Grabens, Canyonlands NP
Arches NP, UT
Basin & Range Province, UT
Squaw Point, OR
Tule Lake, CA
Reverse
Chief Mountain, MT
Banff, Canada
Kananaskis Country, Canada
Strike-Slip
San Andreas Fault, CA
Alpine Fault, New Zealand
Faults & Folds, Makran Range

159

N. Stream Landscapes
Marshak-See For Yourself Sites
    Deep Gorge in the Himalayas
    Headward Erosion, Canyonlands, UT-Overview
    Headward Erosion, Canyonlands, UT-Oblique
    Meanders along Rio Ucayali, Peru
    Incised Meanders, Canyonlands, UT
    Mid-Stream Bars, Rio Negro, Brazil
    Point Bars, Trinity River, TX-Overview
    Point Bars, Trinity River, TX-Oblique
    Trellis Drainage Pattern, PA
    Radial Drainage, Mt. Shasta, CA
    Dendritic Drainage, PA
Stream Features
    Meanders along Rio Ucayali, Peru
    Mid-Stream Bars, Rio Negro, Brazil
    Point Bars, Trinity River, TX
    Mississippi River Delta
Stream "Ages"
    "Youthful"-Montrose, CO
    "Mature"-Leavenworth, KS
    "Old"-Campti, LA
Stream Patterns
    Dendritic-PA
    Dendritic-Mesa Verde NP, CO
    Radial-Navajo Mt., UT
    Radial-Mt. Shasta, CA
    Trellis-Maysville, WV
    Trellis-PA
    Rectangular-Treadway, TN
    Parallel-Colona, CO
    Annular-Maverick Spring, WY
Superposed Stream
    San Rafael Swell, UT
Stream Piracy
    Kaaterskill, NY
Entrenched Meanders
    Goosenecks SP, UT
    Canyonlands NP, UT
    Bowknot Bend, UT
    Strasburg, VA
Stream Terraces
    Otto, WY
    Grass Creek, WY
Headward Erosion
    Canyonlands NP, UT
Channeled Scablands
Deep Gorge in the Himalayas

O. Oceans & Coastlines
Marshak-See For Yourself Sites
    Mid-Atlantic Ridge Bathymetry
    Southern South America Bathymetry-Overview
    Southern South America Bathymetry-Scotia Sea
    Coral Reefs, Pacific-Huahine
    Coral Reefs, Pacific-Tahaa
    Rocky Coast, ME
    Offshore Bar near Cape Hatteras, NC
    Chicago Shoreline
    Fjords of Norway-Overview
    Fjords of Norway-Oblique
    Organic Coast, FL-Overview
    Organic Coast, FL-Oblique
    Sandspit, Cape Cod, MA

Mid-Atlantic Ridge Bathymetry
Southern South America Bathymetry
Coral Reefs, Pacific
Rocky Coast, ME
Offshore Bar near Cape Hatteras, NC
Chicago Shoreline
Fjords of Norway
Organic Coast, FL
Sandspit, Cape Cod, MA
U.S. East Coast Sea Level Changes

P. Groundwater & Karst Landscapes
Marshak-See For Yourself Sites
    Irrigation in the Saudi Desert-Overview
    Irrigation in the Saudi Desert-Detailed
    Water Table, MN
    Everglades, FL
    Hot Springs, Yellowstone NP, WY
    Desert Oasis, Egypt
    Sinking Venice, Italy-Overview
    Sinking Venice, Italy-Detailed
    Sinkholes, FL
    Karst Landscape, Puerto Rico
    Karst Landscape, Puerto Rico-Radio Telescope
    Tower Karst, China
Groundwater
    Everglades, FL
    Irrigation in the Saudi Desert
    Water Table, MN
    Hot Springs, Yellowstone NP, WY
    Desert Oasis, Egypt
    Sinking Venice, Italy
Karst Landscapes
    Sinkholes, FL
    Sinkholes near Interlachen, FL
    Karst Topography near Mammoth Cave NP, KY
    Sinkhole Plain near Avoca, IN
    Karst Topography near Orleans, IN
    Tower Karst, China
    Karst Landscape, Puerto Rico

Q. Desert Landscapes
Marshak-See For Yourself Sites
    Sand Dunes, Namib Desert, Africa
    Uluru (Ayers Rock), Australia
    Uluru (Ayers Rock), Australia-Fold
    Atacama Desert near Chala, Peru
    Tarim Basin, Western China
    Dry Wash, AZ
    Urbanizing a Desert, Tucson, AZ
    Playa in Death Valley, CA
    Buttes, Monument Valley, AZ
    Sand Sea in the Sahara, Egypt
Sand Dunes, Namib Desert, Africa
Uluru (Ayers Rock), Australia
Atacama Desert near Chala, Peru
Tarim Basin, Western China
Dry Wash, AZ
Urbanizing a Desert, Tucson, AZ
Playa in Death Valley, CA
Buttes, Monument Valley, AZ
Sand Sea in the Sahara, Egypt
Aral Sea, Central Asia (Historical Animation)

R. Glacial Landscapes
Marshak-See For Yourself Sites
    Continental Glacier, Antarctica
    Greenland & the Arctic Ocean
    Southern Tip of Greenland
    Baffin Island, Canada-Overview
    Baffin Island, Canada-Valley Glacier
    Matterhorn, Switzerland
    Malaspina Glacier Area, AK-Overview
    Malaspina Glacier Area, AK-Knob & Kettle
    Topography
    Glaciated Peaks, MT-Mt. Cleveland
    Glaciated Peaks, MT-Glacially Scoured Valley
    Sierra Nevada, CA-Upper Sierra
    Sierra Nevada, CA-Yosemite NP
    Pluvial Lake Shore, Salt Lake City, UT
Continental Glaciers
    Continental Glacier, Antarctica
    Greenland
    Malaspina Glacier Area, AK
    Glacial Moraines-Long Island, NY & Cape Cod, MA
    Drumlins/Flutes, Palmyra, NY
    Pluvial Lake Shore, Salt Lake City, UT
Alpine/Valley Glaciers
    Baffin Island, Canada
    Matterhorn, Switzerland
    Sierra Nevada, CA
    Hanging Valley, Bridalveil Falls, CA
    Roche Moutonnee-Lembert Dome, CA
    Lateral Moraines-Mono Lake, CA
    End Moraine, Holy Cross, CO
    Rocky Mountain NP, CO
    Glaciated Peaks, MT
    Glacier National Park, MT

S. Global Change
Marshak-See For Yourself Sites
    Clear Cutting the Amazon, Brazil-Overview
    Clear Cutting the Amazon, Brazil-Detailed
    Long-Term Deforestation, Brazil-Barren Hills
    Long-Term Deforestation, Brazil-Brazilian Highlands
    Edge of Everglades, FL-Transition
    Edge of Everglades, FL-Concrete "Forest"
    Urbanizing the Desert, AZ
    Receding Glacier, Switzerland
    Melting Permafrost, Siberia
    Intense Urbanization, Tokyo, Japan
    Village and Fields, China
    Desertification of the Sahel, Africa
Deforestation
    Clear Cutting the Amazon, Brazil
    Long-Term Deforestation, Brazil
    Deforestation, South America (Historical Animation)
Melting Ice
    Larsen B Ice Shelf, Antarctica (Historical Animation)
    Receding Glacier, Switzerland
    Melting Permafrost, Siberia
Urbanization
    Village and Fields, China
    Intense Urbanization, Tokyo, Japan
    Urbanizing the Desert, AZ
    Everglades, FL
Desertification of the Sahel, Africa
Köppen-Geiger Climate Classification